Pulmonary Physiology

Pulmonary Physiology

Michael G. Levitzky, Ph.D.
Associate Professor of Physiology
Louisiana State University Medical Center

McGraw-Hill Book Company

New York St. Louis San Francisco Auckland Bogotá Guatemala
Hamburg Johannesburg Lisbon London Madrid Mexico Montreal
New Delhi Panama Paris San Juan São Paulo Singapore Sydney
Tokyo Toronto

PULMONARY PHYSIOLOGY

3 4 5 6 7 8 9 0 DODO 8 9 8 7 6 5 4

ISBN 0-07-037431-7

This book was set in Times Roman by Bi-Comp, Incorporated. The editors were Anna R. Ferrera and John J. Fitzpatrick; the production supervisor was Jeanne Skahan; the designer was Joe Cupani.
R. R. Donnelley & Sons Company was printer and binder.

Library of Congress Cataloging in Publication Data

Levitzky, Michael G.
Pulmonary physiology.

Bibliography: p.
Includes index.
1. Lungs. 2. Respiration. I. Title.
QP121.L45 612'.2 81-17127
ISBN 0-07-037431-7 AACR2

To Robert S. Alexander

Contents

Preface

My goal in writing this book is to provide the first-year medical student (as well as respiratory therapy, nurse-anesthesia, and other students) with a complete and solid background in the aspects of pulmonary physiology essential for an understanding of clinical medicine. My approach is to encourage self-sufficiency not only in studying pulmonary physiology for the first time but also in understanding the basic concepts of pulmonary physiology well enough to apply them with confidence to future patients. I believe that the ways to accomplish this are: to inform the student of the goals of each chapter with clearly stated learning objectives, to give detailed and complete *explanations* of physiological mechanisms and demonstrate how they apply to pathological states, and to give the student a means of self-testing by providing study questions and problems.

The challenge is to write a book that students can read without difficulty in the limited amount of time allocated to pulmonary physiology in the typical first-year curriculum. The material must be presented in a way that discourages memorization without real comprehension because only those students who understand the basic mechanisms are able to apply them to new situations. The result of this approach should be a book that covers the essentials of the respiratory system as concisely as possible yet raises no questions in the student's mind without answering them. I hope that I have achieved these goals in writing this book.

I take this opportunity to thank those scientists who helped to shape my career as a physiologist, including Drs. John A. Krasney, Robert E. Dutton, Jonathan C. Newell, and the late Samuel R. Powers, Jr. I also acknowledge the assistance of those who helped me prepare this book, including Drs. Stanley M. Hall and Remi Gomila, and also Charles Chapman, Virginia Howard, Cynthia Sims of the Louisiana State University Medical Center Editorial Office, and my wife, Ellen.

Michael G. Levitzky

Pulmonary Physiology

Chapter 1

Function and Structure of the Respiratory System

OBJECTIVES

The student states the functions of the respiratory system and relates the structural organization of the system to its functions.

1 Describes the exchange of oxygen and carbon dioxide with the atmosphere and relates gas exchange to the metabolism of the tissues of the body.
2 Defines the role of the respiratory system in acid-base balance.
3 Lists the nonrespiratory functions of the lungs.
4 Defines and describes the alveolar-capillary unit, the site of gas exchange in the lungs.
5 Describes the transport of gas through the conducting airways to and from the alveoli.
6 Describes the structural characteristics of the airways.
7 Lists the components of the chest wall and relates the functions of the muscles of respiration to the movement of air into and out of the alveoli.
8 Describes the central nervous system initiation of breathing and the innervation of the respiratory muscles.

Most of the tissues of the body require oxygen to produce energy, and a continuous supply of oxygen is necessary for the normal function of these tissues. Carbon dioxide is a by-product of this aerobic metabolism, and it must be removed from the vicinity of the metabolizing cells. The main functions of the respiratory system are to obtain oxygen from the external environment to supply it to the cells, and to remove from the body the carbon dioxide produced by cellular metabolism.

The *respiratory system* is composed of the lungs, the conducting airways, the parts of the central nervous system concerned with the control of the muscles of respiration, and the chest wall. The chest wall consists of the muscles of respiration, such as the diaphragm and the intercostal muscles, and the rib cage.

FUNCTIONS OF THE RESPIRATORY SYSTEM

The functions of the respiratory system include gas exchange, acid-base balance, phonation, pulmonary defense and metabolism, and the handling of bioactive materials.

Gas Exchange

The exchange of carbon dioxide for oxygen takes place in the lungs. Fresh air, containing oxygen, is inspired into the lungs through the conducting airways by forces generated by the respiratory muscles, acting on commands initiated by the central nervous system. At the same time, mixed venous blood from the various body tissues, which has a high content of carbon dioxide and a low content of oxygen, is pumped into the lungs by the right ventricle of the heart. In the pulmonary capillaries, carbon dioxide is exchanged for oxygen and the blood leaving the lungs, which now has a high oxygen content and a relatively low carbon dioxide content, is distributed to the tissues of the body by the left side of the heart. During expiration, gas with a high concentration of carbon dioxide is expelled from the body. A schematic diagram of the gas exchange function of the respiratory system is shown in Fig. 1-1.

Other Functions

Acid-Base Balance In the body, increases in carbon dioxide lead to increased hydrogen ion concentration because of the following reaction:

$$CO_2 + H_2O \rightleftharpoons H_2CO_3 \rightleftharpoons H^+ + HCO_3^-$$

The respiratory system can therefore participate in acid-base balance by removing CO_2 from the body. The central nervous system has sensors for the CO_2 levels in the arterial blood and in the cerebrospinal fluid,

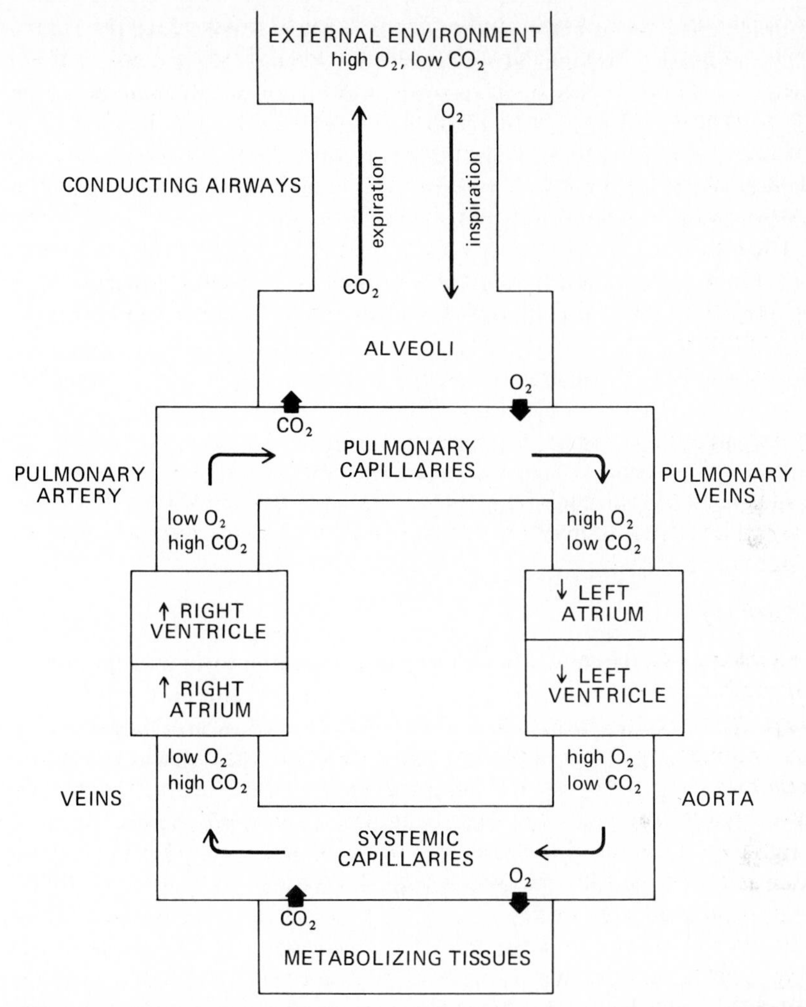

Figure 1-1 Schematic representation of gas exchange between the tissues of the body and the environment.

which send information to the controllers of breathing. Acid-base balance is discussed in greater detail in Chap. 8; the control of breathing is discussed in Chap. 9.

Phonation Phonation is the production of sounds by the movement of air through the vocal cords. Speech, singing, and other sounds are produced by the action of the central nervous system controllers on the muscles of respiration, causing air to flow through the vocal cords. Phonation will not be discussed in detail in this book.

Pulmonary Defense Mechanisms Each breath brings into the lungs a small sample of the local atmospheric environment. This may include microorganisms such as bacteria, dust, particles of silica or asbestos, toxic gases, smoke (cigarette and other types), and other pollutants. In addition, the temperature and humidity of the local atmosphere vary tremendously. The mechanisms by which the lungs are protected from these environmental assaults are discussed in Chap. 10.

Pulmonary Metabolism and the Handling of Bioactive Materials The cells of the lung must metabolize substrate to supply energy and nutrients for their own maintenance. Some specialized pulmonary cells also produce substances necessary for normal pulmonary function. In addition, the pulmonary capillary endothelium contains a great number of enzymes that can metabolize or modify naturally occurring vasoactive substances found in venous blood. These metabolic functions of the respiratory system are discussed in Chap. 10.

STRUCTURE OF THE RESPIRATORY SYSTEM

Air enters the respiratory system through the nose or mouth. Air entering through the nose is filtered, heated to body temperature, and humidified as it passes through the nose and nasal turbinates. These protective mechanisms are discussed in Chap. 10. The upper airways are shown in Fig. 10-1. Air breathed through the nose enters the airways via the nasopharynx, and through the mouth via the oropharynx. It then passes through the glottis and the larynx and enters the tracheobronchial tree. After passing through the conducting airways, the inspired air enters the alveoli, where it comes into intimate contact with the mixed venous blood in the pulmonary capillaries.

The Alveolar-Capillary Unit

The *alveolar-capillary* unit is the site of gas exchange in the lung. The alveoli, estimated to number about 300 million, are almost completely enveloped in pulmonary capillaries. There may be as many as 280 billion pulmonary capillaries, or nearly 1000 pulmonary capillaries per alveolus. The result of these staggering numbers of alveoli and pulmonary capillaries is a vast area of contact between alveoli and capillaries—probably 50 to 100 m^2 of common surface area available for gas exchange by diffusion. The alveoli are about 250 μm in diameter.

Figure 1-2 is a scanning electron micrograph of the alveolar-capillary surface. Figure 1-3 shows an even greater magnification of the site of gas exchange.

The alveolar septum appears to be almost entirely composed of pulmonary capillaries. Red blood cells (erythrocytes) can be seen inside the

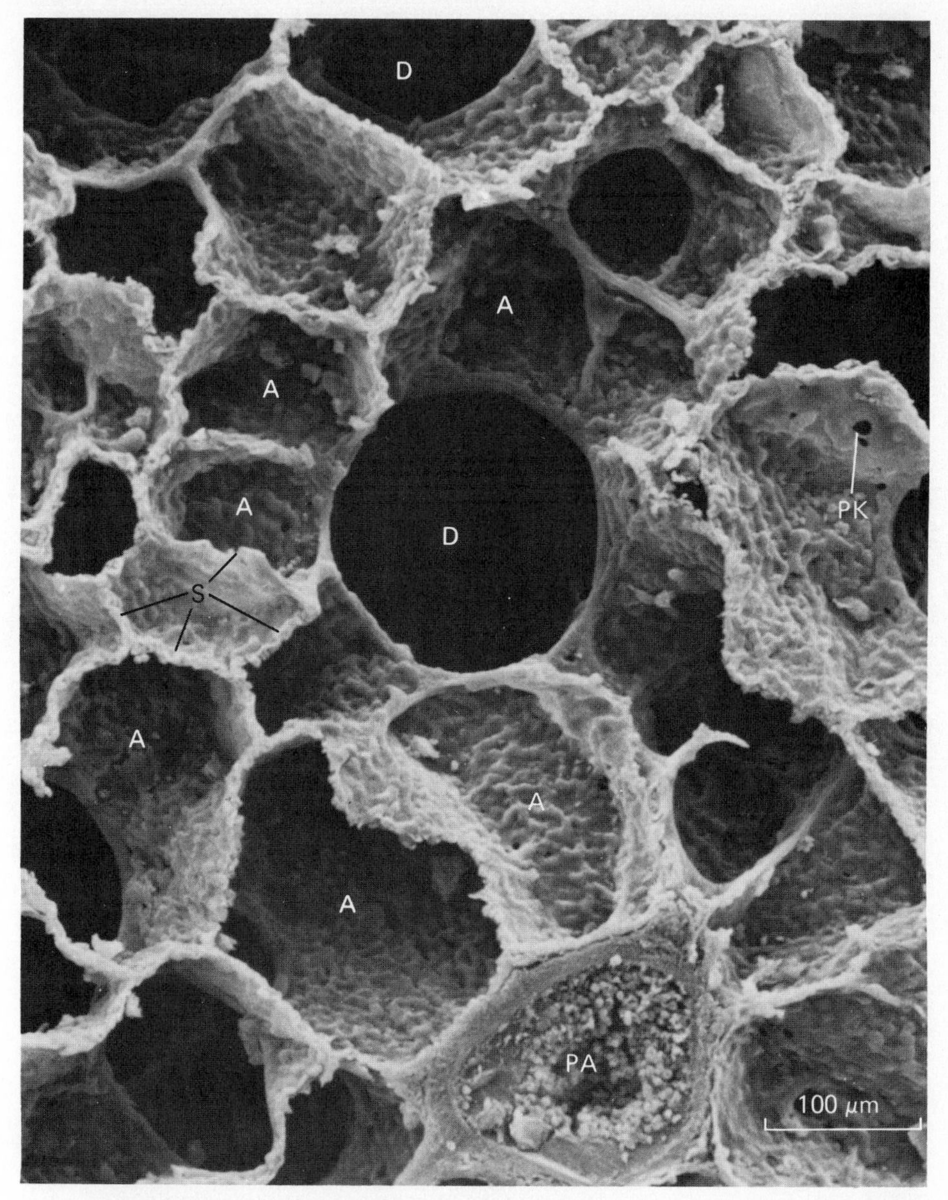

Figure 1-2 Scanning electron micrograph of human lung parenchyma. A = Alveoli; S = Alveolar septa; D = Alveolar duct; PK = Pore of Kohn; PA = Small branch of the pulmonary artery. (*Reproduced with permission from Weibel, 1980.*)

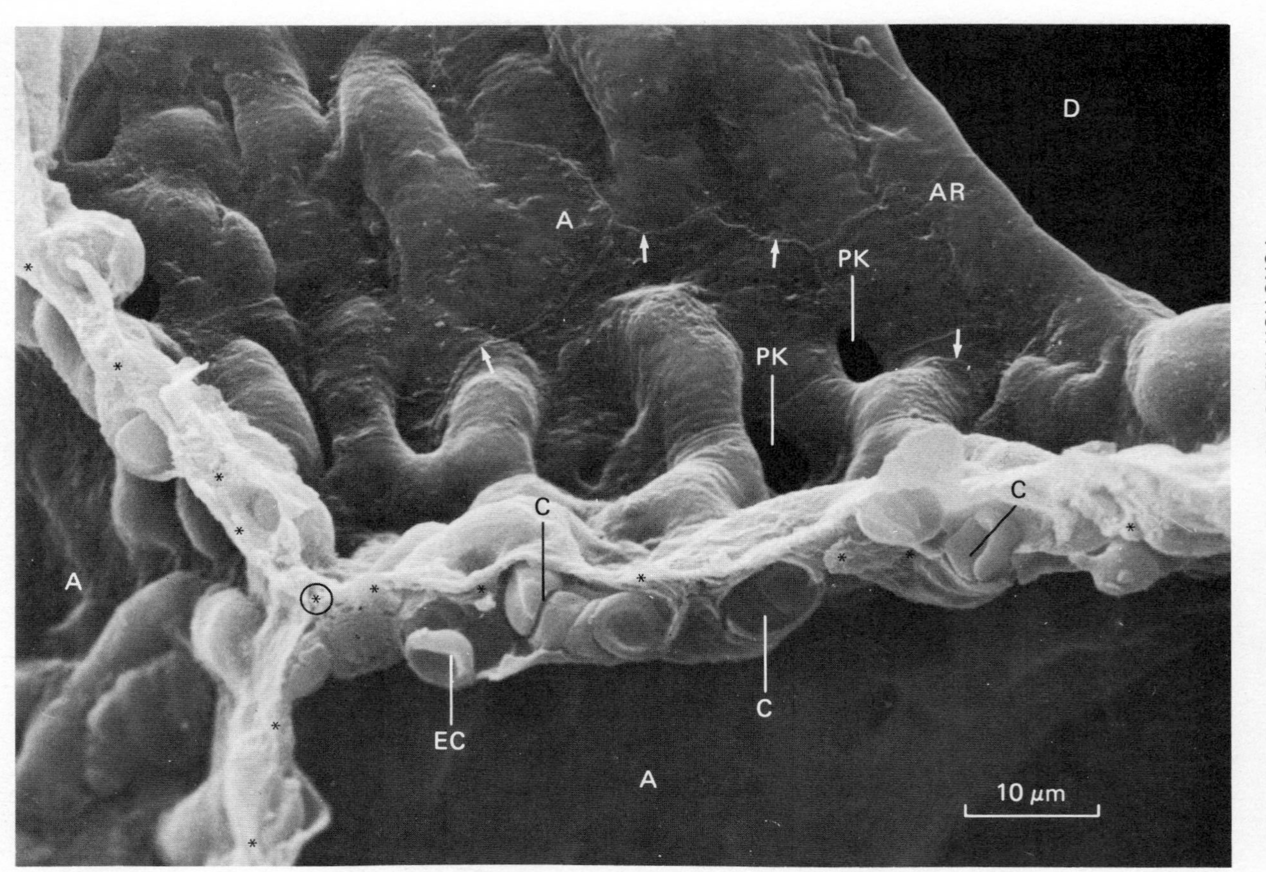

Figure 1-3 Scanning electron micrograph of the surface and cross section of an alveolar septum. Capillaries (C) are seen sectioned in the foreground, with erythrocytes (EC) within them. A = Alveolus; D = Alveolar duct; PK = Pore of Kohn; AR = Alveolar entrance to duct; * = Connective tissue fibers. The encircled asterisk is at a junction of

capillaries at the point of section. Elastic and connective tissue fibers, not visible in the figure, are found between the capillaries in the alveolar septa. Also shown in this figure are the *pores of Kohn* or interalveolar communications.

The Alveolar Surface The alveolar surface is mainly composed of a single thin layer of squamous epithelial cells, the *type I alveolar cells.* Interspersed among these are the larger cuboidal *type II alveolar cells,* which produce the fluid layer that lines the alveoli. The alveolar surface fluid layer is discussed in detail in Chap. 2. A third cell type, the free-ranging phagocytic *alveolar macrophage,* is found in varying numbers in the extracellular lining of the alveolar surface. These cells patrol the alveolar surface and phagocytize inspired particles such as bacteria. Their function is discussed in Chap. 10.

The Capillary Endothelium A cross section of a pulmonary capillary is shown in the transmission electron micrograph in Fig. 1-4. An erythrocyte is seen in cross section in the lumen of the capillary. Capillaries are formed by a single layer of squamous epithelial cells that are aligned to form tubes. The nucleus of one of the capillary endothelial cells can be seen in the micrograph.

The barrier to gas exchange between the alveoli and pulmonary capillaries can also be seen in Fig. 1-4. It consists of the alveolar epithelium, the capillary endothelium, and the interstitial space between them. Gases must also pass through the fluid lining the alveolar surface (not visible in Fig. 1-4) and the plasma in the capillary. The barrier to diffusion is about 0.5 μm thick. Gas exchange by diffusion is discussed in Chap. 6.

The Airways

After passing through the nose or mouth, pharynx, and larynx (*the upper airways*), air enters the tracheobronchial tree. Starting with the trachea, the air may pass through as few as 10 or as many as 23 *generations,* or branchings, on its way to the alveoli. The branchings of the tracheobronchial tree and its nomenclature are shown in Fig. 1-5. Alveolar gas exchange units are denoted by the U-shaped sacs.

The first 16 generations of airways, the *conducting zone,* contain no alveoli and thus are anatomically incapable of gas exchange with the venous blood. They constitute the *anatomic dead space,* which is discussed in Chap. 3. Alveoli start to appear at the seventeenth through the nineteenth generations, in the respiratory bronchioles, which constitute the *transitional zone.* The twentieth to twenty-second generations are lined with alveoli. These *alveolar ducts* and the *alveolar sacs,* which terminate the tracheobronchial tree, are referred to as the *respiratory zone.*

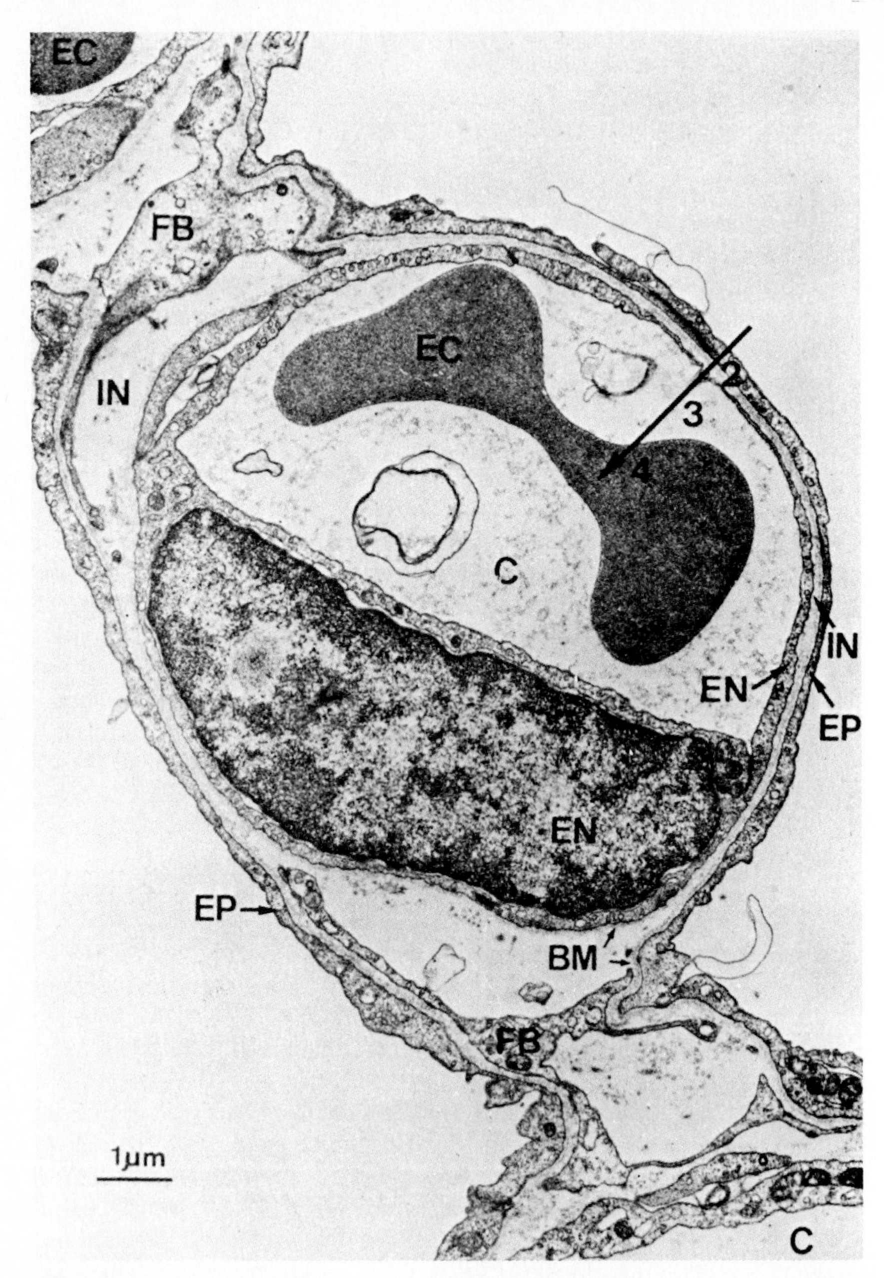

Figure 1-4 Transmission electron micrograph of a cross section of a pulmonary capillary. An erythrocyte (EC) is seen within the capillary. C = Capillary; EN = Capillary endothelial cell (note its large nucleus); EP = Alveolar epithelial cell; IN = interstitial space; BM = Basement membrane; FB = Fibroblast processes; FC = Connective tissue fibrils; 2,3,4 = Diffusion pathway through the alveolar-capillary barrier, the plasma, and the erythrocyte, respectively. (*Reproduced with permission from Weibel, 1970.*)

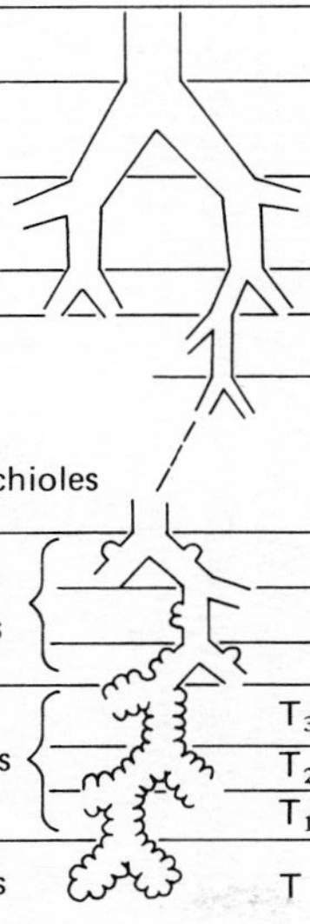

Zone	Generation			Diameter, cm	Length, cm	Number	Total cross sectional area, cm^2
conducting zone	trachea		0	1.80	12.0	1	2.54
	bronchi		1	1.22	4.8	2	2.33
	↓		2	0.83	1.9	4	2.13
			3	0.56	0.8	8	2.00
	bronchioles		4	0.45	1.3	16	2.48
	↓		5	0.35	1.07	32	3.11
			↓	↓	↓	↓	↓
	terminal bronchioles		16	0.06	0.17	6×10^4	180.0
transitional and respiratory zones	respiratory bronchioles		17	↓	↓	↓	↓
			18				
			19	0.05	0.10	5×10^5	10^3
	alveolar ducts	T_3	20	↓	↓	↓	↓
		T_2	21				
		T_1	22				
	alveolar sacs	T	23	0.04	0.05	8×10^6	10^4

Figure 1-5 Schematic representation of airway branching in the human lung with approximate dimensions. (*Figure after Weibel, 1963. Data from Bouhuys, 1977. Reproduced with permission.*)

The portion of lung supplied by a primary respiratory bronchiole is called an *acinus*. All of the airways of an acinus participate in gas exchange. The numerous branchings of the airways result in a tremendous total cross-sectional area of the distal portions of the tracheobronchial tree, even though the diameters of the individual airways are quite small. This can be seen in the table accompanying Fig. 1-5.

Structure of the Airways The structure of the airways varies considerably, depending on their location in the tracheobronchial tree. The trachea is a fibromuscular tube supported ventrolaterally by C-shaped cartilage and completed dorsally by smooth muscle. The cartilage of the large bronchi is semicircular, like that of the trachea, but as the bronchi enter the lungs, the cartilage rings disappear and are replaced by irregularly shaped cartilage plates. They completely surround the bronchi and give the intrapulmonary bronchi their cylindrical shape. These plates, which help support the larger airways, diminish progressively in the distal airways and disappear in airways about 1 mm in diameter. By definition, airways with no cartilage are termed *bronchioles*. Because the bronchioles and alveolar ducts contain no cartilage support, they are subject to collapse when compressed. This tendency is partly opposed by the attachment of the alveolar septa, containing elastic tissue, to their walls, as seen in Fig. 1-2 and shown schematically in Fig. 2-18. As the cartilage plates become irregularly distributed around distal airways, the muscular layer completely surrounds these airways. The muscular layer is intermingled with elastic fibers. As the bronchioles proceed toward the alveoli, the muscle layer becomes thinner, although smooth muscle can even be found in the walls of the alveolar ducts. The outermost layer of the bronchiolar wall is surrounded by dense connective tissue with many elastic fibers.

The Lining of the Airways The entire respiratory tract, except for part of the pharynx, the anterior third of the nose, and the terminal respiratory units distal to the respiratory bronchioles, is lined with ciliated cells interspersed with mucus-secreting goblet cells and other secretory cells. The ciliated cells are pseudostratified columnar cells in the larger airways and become cuboidal in the bronchioles. In the bronchioles the goblet cells become less frequent and are replaced by another type of secretory cell, the Clara cell. The ciliated epithelium, along with the mucus secreted by glands along the airways and the goblet cells and the secretory products of the Clara cells, constitutes an important mechanism for the protection of the lung. This mechanism is discussed in detail in Chap. 10.

The Muscles of Respiration and the Chest Wall

The muscles of respiration and the chest wall are essential components of the respiratory system. The lungs are not capable of inflating themselves—the force for this inflation must be supplied by the muscles of respiration. The chest wall must be intact and able to expand if air is to enter the alveoli. The interactions between the muscles of respiration and the chest wall and the lungs are discussed in detail in the next chapter.

The primary components of the chest wall are shown schematically in Fig. 1-6. These include the rib cage; the external and internal intercostal muscles and the diaphragm, which are the main muscles of respiration; and the lining of the chest wall, the visceral and parietal pleura. Other muscles of respiration include the abdominal muscles, including the rectus abdominis; the parasternal intercartilaginous muscles; and the accessory muscles of inspiration, the sternocleidomastoid and scalenus muscles.

The Central Nervous System and Neural Pathways

Another important component of the respiratory system is the central nervous system. Unlike cardiac muscle, the muscles of respiration do not contract spontaneously. Each breath is initiated in the brain, and this message is carried to the respiratory muscles via the spinal cord and the nerves innervating the respiratory muscles.

Figure 1-6 The primary components of the chest wall.

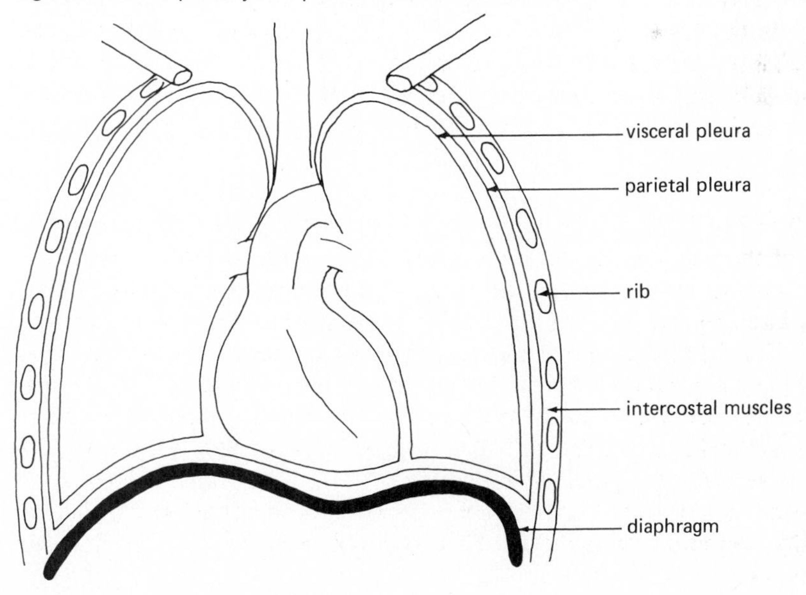

Spontaneous automatic breathing is generated by groups of neurons located in the medulla. This medullary respiratory center is also the final integration point for influences from higher brain centers, such as those concerned with voluntary breathing and speech; for information from chemoreceptors in the blood and cerebrospinal fluid; and for afferent information from neural receptors in the airways, joints and muscles, skin, and elsewhere in the body. The control of breathing is discussed in Chap. 9.

Chapter 2

Mechanics of Breathing

OBJECTIVES

The student understands the mechanical properties of the lung and chest wall during breathing.

1. Describes the generation of a pressure gradient between the atmosphere and the alveoli.
2. Describes the passive expansion and recoil of the alveoli.
3. Defines the mechanical interaction of the lung and chest wall and relates this concept to the negative intrapleural pressure.
4. Describes the pressure-volume characteristics of the lung and chest wall and predicts changes in the compliance of the lung and chest wall in different physiological and pathological conditions.
5. States the roles of pulmonary surfactant and alveolar interdependence in the recoil and expansion of the lung.
6. Defines the functional residual capacity (FRC) and uses his or her understanding of lung–chest wall interactions to predict changes in FRC in different physiological and pathological conditions.
7. Defines airways resistance and lists the factors that contribute to or alter the resistance to airflow.

8 Describes the dynamic compression of airways during a forced expiration.
9 Relates changes in the dynamic compliance of the lung to alterations in airways resistance.
10 Lists the factors that contribute to the work of breathing.
11 Predicts alterations in the work of breathing in different physiological and pathological states.

Air, like other fluids, moves from a region of higher pressure to one of lower pressure. Therefore, for air to be moved into or out of the lungs, a pressure difference between the atmosphere and the alveoli must be established. If there is no pressure gradient, no air flow will occur.

Under normal circumstances, inspiration is accomplished by causing alveolar pressure to fall below atmospheric pressure. Because, for the purpose of discussing the mechanics of breathing, atmospheric pressure is conventionally referred to as 0 cmH_2O, lowering alveolar pressure below atmospheric pressure is known as *negative-pressure breathing*. As soon as a pressure gradient sufficient to overcome the resistance to airflow offered by the conducting airways is established between the atmosphere and the alveoli, air flows into the lungs. It is also possible to cause air to flow into the lungs by an an apparatus that mechanically raises the pressure at the nose and mouth above alveolar pressure. This *positive-pressure breathing* is generally used on patients unable to generate a sufficient pressure gradient between the atmosphere and the alveoli by normal negative-pressure breathing. Air flows out when alveolar pressure is sufficiently greater than atmospheric pressure to overcome the resistance to airflow offered by the conducting airways.

GENERATION OF A PRESSURE GRADIENT BETWEEN ATMOSPHERE AND ALVEOLI

During normal negative-pressure breathing, alveolar pressure is made lower than that of the atmosphere. This is accomplished by causing the muscles of inspiration to contract, which increases the volume of the thoracic cavity, lowering the intrapleural pressure and expanding the alveoli.

Passive Expansion of Alveoli

The alveoli are not capable of expanding themselves. They only expand passively in response to an increased distending pressure across the alveolar wall. This increased *transmural pressure gradient,* generated by the muscles of inspiration, opens the highly distensible alveoli and thus lowers the alveolar pressure. The transmural pressure gradient is conventionally calculated by subtracting the outside pressure (in this case, the intrapleural pressure) from the inside pressure (in this case, the alveolar pressure).

Negative Intrapleural Pressure

The pressure in the pleural space outside the lung is normally slightly subatmospheric, even when no inspiratory muscles are contracting. This *negative intrapleural pressure* (sometimes also referred to as *negative intrathoracic pressure*) of -3 to -5 cmH_2O is mainly caused by the mechanical interaction between the lung and the chest wall. At the end of expiration, when all of the respiratory muscles are relaxed, the lung and chest wall are acting on each other in opposite directions. The lung is tending to *decrease* its volume because of the *inward* elastic recoil of the distended alveolar walls; the chest wall is tending to *increase* its volume because of its *outward* elastic recoil. Thus the chest wall is acting to hold the alveoli open in opposition to their elastic recoil. Similarly the lung is acting, by its elastic recoil, to hold the chest wall in. Because of this interaction, the pressure is negative at the surface of the very thin (about 5 to 10 μm in thickness at normal lung volumes, with a total volume of about 2 ml), fluid-filled pleural space, as seen at left in Fig. 2-1. There is normally no gas in the intrapleural space, and the lung is held against the chest wall by the thin layer of serous intrapleural liquid.

Initially, before any airflow occurs, the pressure inside the alveoli is the same as atmospheric pressure—by convention 0 cmH_2O. The muscles of inspiration act to increase the volume of the thoracic cavity. As the inspiratory muscles contract, expanding the thoracic volume and increasing the outward stress on the lung, the intrapleural pressure becomes *more negative*. Therefore the transmural pressure gradient tending to distend the alveolar wall, sometimes called the *transpulmonary pressure,* increases as shown in Fig. 2-1, and the alveoli enlarge passively, lowering alveolar pressure and establishing the pressure gradient for airflow into the lung. In reality, only a small percentage of the total number of alveoli are directly exposed to the intrapleural surface pressure, and at first thought it is difficult to see how alveoli located centrally in the lung could be expanded by a more negative intrapleural pressure. Careful analysis has shown, however, that the pressure at the pleural surface is transmitted through the alveolar walls to more centrally located alveoli. This structural *interdependence* of alveolar units is demonstrated in Fig. 2-2.

The Muscles of Respiration

Inspiratory Muscles The muscles of inspiration include the diaphragm, the external intercostals, and the accessory muscles of inspiration.

The Diaphragm The diaphragm is a large dome-shaped muscle (about 250 cm^2 in surface area) that separates the thorax from the abdominal cavity. As mentioned in Chap. 1, the diaphragm is considered to be an

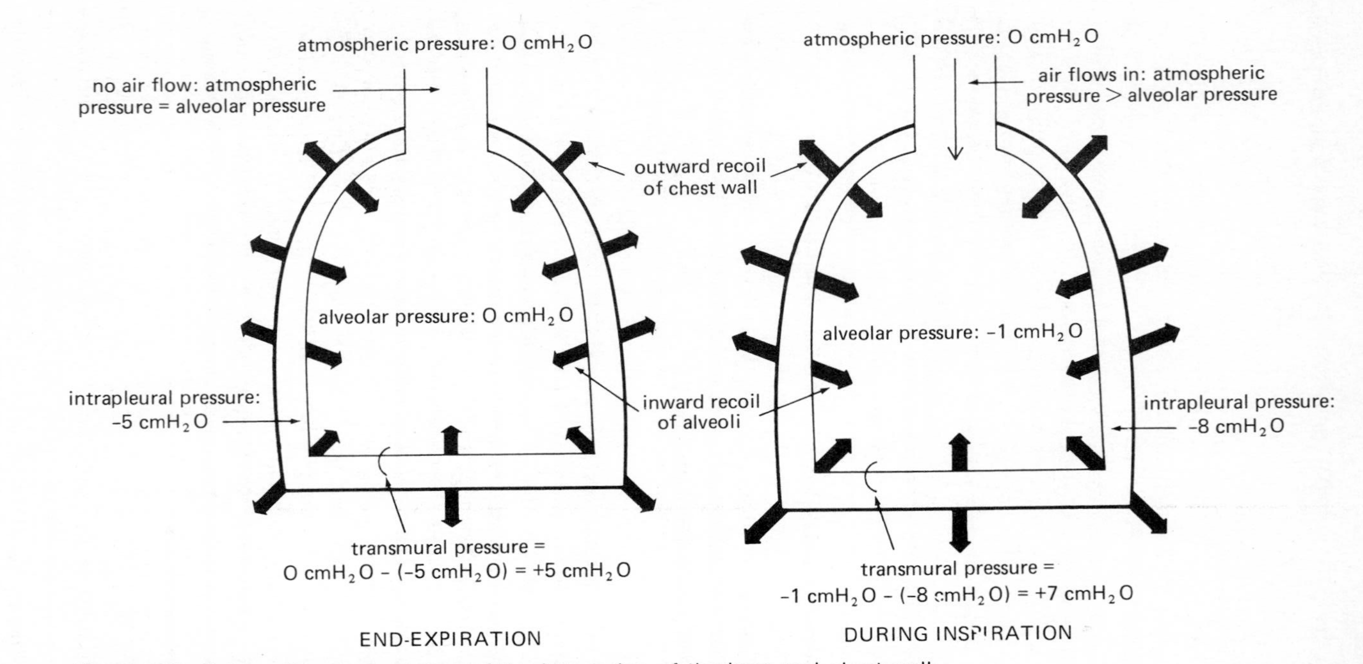

Figure 2-1 Schematic representation of the interaction of the lung and chest wall. *Left:* At end-expiration the muscles of respiration are relaxed. The inward elastic recoil of the lung is balanced by the outward elastic recoil of the chest wall. Intrapleural pressure is −5 cmH$_2$O; alveolar pressure is 0. The transmural pressure gradient across the alveolus is therefore 0 cmH$_2$O − (−5 cmH$_2$O), or 5 cmH$_2$O. Since alveolar pressure is equal to atmospheric pressure, no airflow occurs.
Right: During inspiration, contraction of the muscles of inspiration causes intrapleural pressure to become more negative. The transmural pressure gradient increases and the alveoli are distended, decreasing alveolar pressure below atmospheric pressure, which causes air to flow into the alveoli.

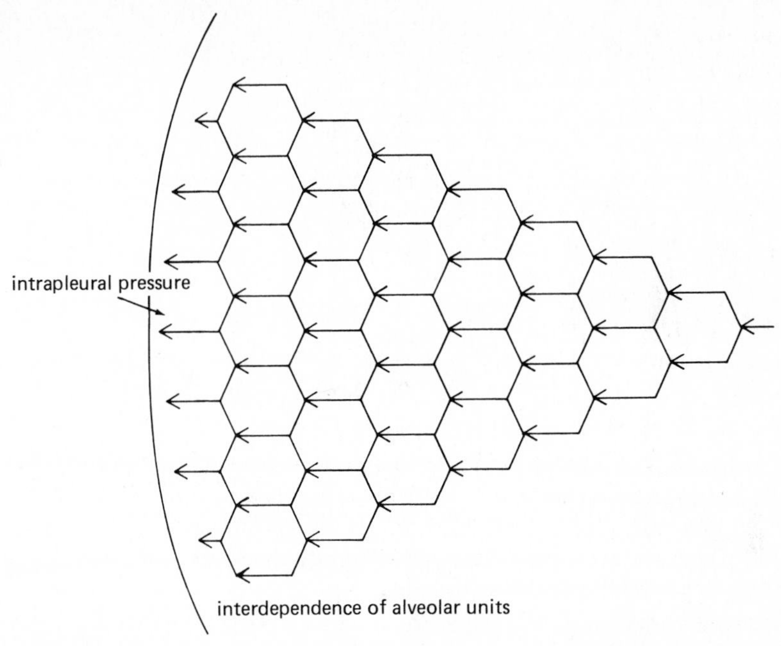

Figure 2-2 Structural interdependence of alveolar units. The pressure gradient across the outermost alveoli is transmitted mechanically through the lung via the alveolar septa.

integral part of the chest wall and must always be considered in the analysis of chest wall mechanics.

The diaphragm is innervated by the two phrenic nerves, which leave the spinal cord at the third through the fifth cervical segments. When the diaphragm contracts, its dome descends into the abdominal cavity, elongating the thorax. Because it is inserted into the lower rib margins, the lower ribs are also elevated during diaphragmatic contraction, as shown in Fig. 2-3.

The diaphragm is the primary muscle of inspiration and is responsible for about two-thirds of the air that enters the lungs during normal, quiet breathing (which is called *eupnea*). The dome of the diaphragm moves 1 to 2 cm downward into the abdominal cavity during normal, quiet breathing; it can descend as much as 10 cm during a deep inspiration. If one of the leaflets of the diaphragm is paralyzed, for example, because of transection of one of the phrenic nerves, it will "paradoxically" move up into the thorax as intrapleural pressure becomes more negative during a rapid inspiratory effort.

The External Intercostals When they are stimulated to contract, the external intercostals (and parasternal intercartilaginous muscles) raise and enlarge the rib cage. Figure 2-4 demonstrates that this action in-

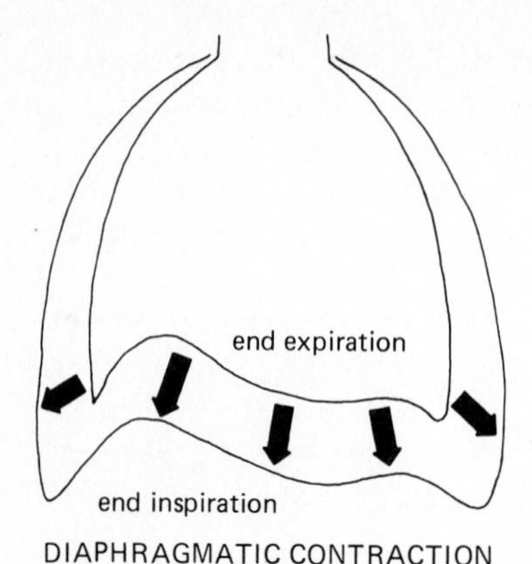

Figure 2-3 Illustration of the actions of diaphragmatic contraction in expanding the thoracic cavity.

creases the anterioposterior dimension of the chest as the ribs rotate upward about their axes (the so-called bucket-handle motion) and also increases the transverse dimension of the lower portion of the chest. These muscles are innervated by nerves leaving the spinal cord at the first through the eleventh thoracic segments.

The Accessory Muscles of Inspiration The accessory muscles of inspiration are not involved during normal quiet breathing but may be called into play during exercise, during the inspiratory phase of coughing or sneezing, or in a pathological state, such as asthma. The *scalene* muscles help to enlarge the upper rib cage. The *sternocleidomastoid* elevates the sternum and helps to increase the anterioposterior and transverse dimensions of the chest.

Figure 2-4 Illustration of the actions of contraction of the external intercostals in expanding and elevating the rib cage.

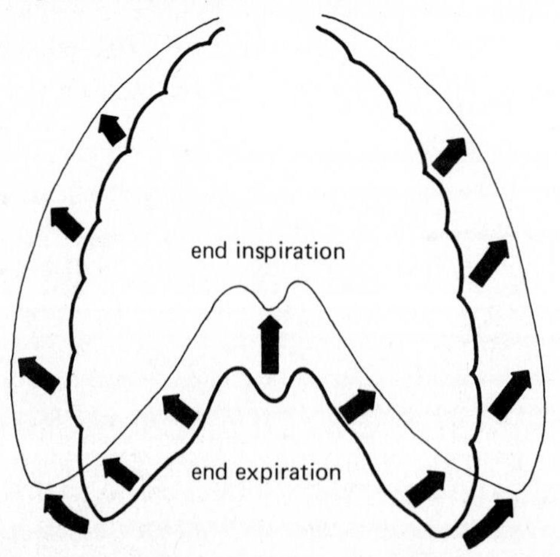

Expiratory Muscles Expiration is *passive* during normal quiet breathing and no respiratory muscles contract. As the inspiratory muscles relax, the increased elastic recoil of the distended alveoli is sufficient to decrease the alveolar volume and raise alveolar pressure above atmospheric pressure. Now the pressure gradient for airflow out of the lung has been established.

Active expiration occurs during exercise, speech, singing, the expiratory phase of coughing or sneezing, and in a pathological state such as chronic bronchitis. The muscles of expiration are *the muscles of the abdominal wall,* including the rectus abdominis, the external and internal oblique muscles, the transversus abdominis, and the *internal intercostal muscles*.

The Abdominal Muscles When the abdominal muscles contract, they compress the abdominal contents against the relaxed diaphragm, forcing it upward into the thoracic cavity. They also help to depress the lower ribs and pull down the anterior part of the lower chest.

The Internal Intercostal Muscles Contraction of the internal intercostal muscles depresses the rib cage downward in a manner opposite to the actions of the external intercostals.

Summary of the Events Occurring During the Course of a Breath

The events occurring during the course of an idealized normal quiet breath are shown in Fig. 2-5. For the purpose of clarity, inspiration and expiration are considered to be of equal duration.

The *volume* of air entering or leaving the lungs can be measured with a spirometer, as will be described in Chap. 3. *Airflow* can be measured by breathing through a pneumotachograph, which measures the pressure differential across a fixed resistance. The *intrapleural pressure* can be estimated by having a subject swallow a balloon into the intrathoracic portion of the esophagus. The pressure then measured in the balloon is nearly equal to intrapleural pressure. *Alveolar pressures* are not directly measurable and must be calculated.

Initially, alveolar pressure equals atmospheric pressure so that no air flows into the lung. Intrapleural pressure is -5 cmH_2O. Contraction of the inspiratory muscles causes intrapleural pressure to become more negative and the alveoli are distended. Note the two different courses for changes in intrapleural pressure. The *dashed line* predicts the changes in intrapleural pressure necessary to overcome the elastic recoil of the alveoli. The *solid line* is a more accurate representation of intrapleural pressure because it also includes the additional pressure work that must be done to overcome the resistance to airflow and tissue resistance discussed later in this chapter. As the alveoli are distended, the pressure inside them falls below atmospheric pressure and air flows into the alveoli, as seen in the

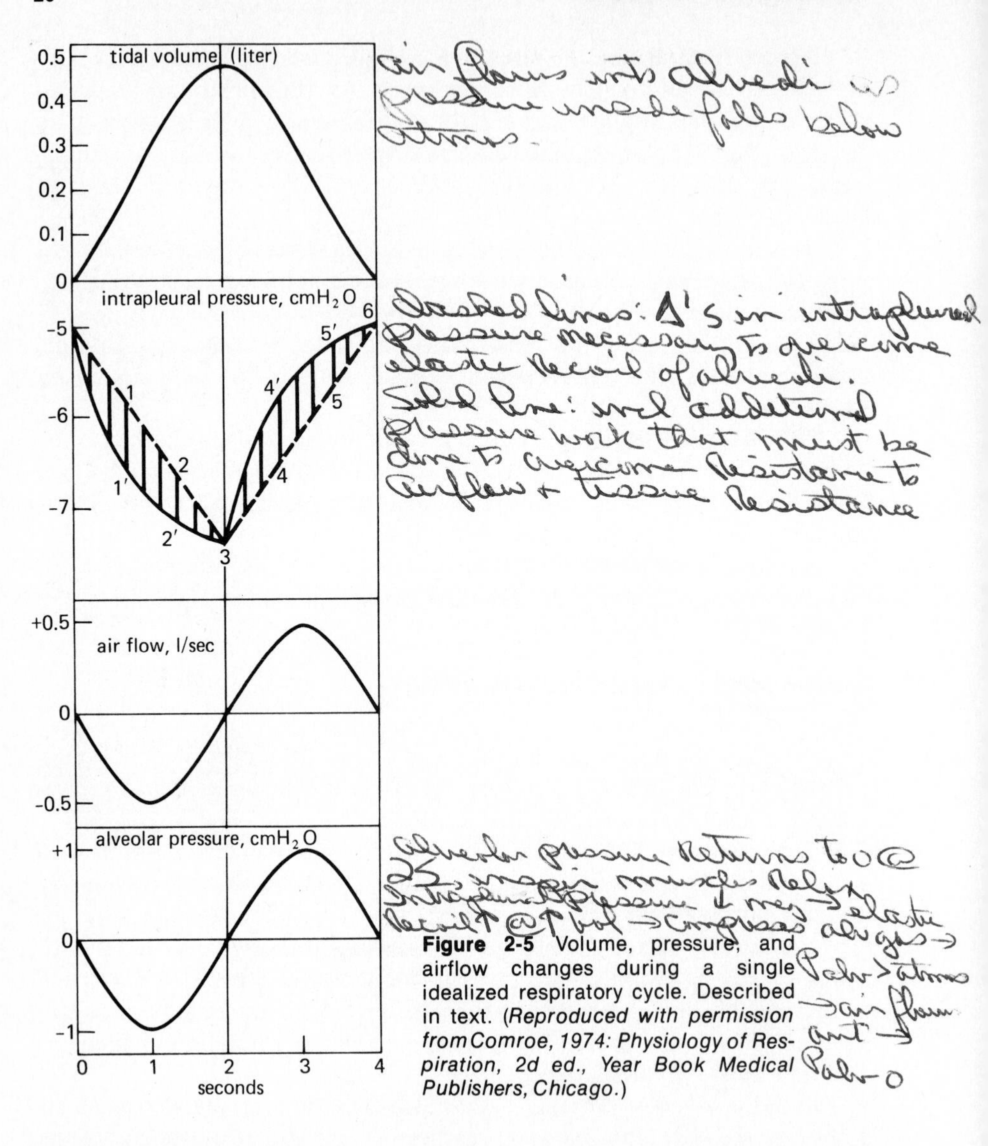

Figure 2-5 Volume, pressure, and airflow changes during a single idealized respiratory cycle. Described in text. (*Reproduced with permission from Comroe, 1974: Physiology of Respiration, 2d ed., Year Book Medical Publishers, Chicago.*)

tidal volume panel. As the air flows into the alveoli, alveolar pressure returns to 0 cmH_2O and airflow into the lung ceases. At the vertical line (at 2 s) the inspiratory effort ceases and the inspiratory muscles relax. Intrapleural pressure becomes less negative and the elastic recoil of the alveolar walls, which is increased at the higher lung volume, is allowed to compress the alveolar gas. This process raises alveolar pressure above atmospheric pressure so that air flows out of the lung until an alveolar pressure of 0 cmH_2O is restored.

PRESSURE-VOLUME RELATIONSHIPS IN THE RESPIRATORY SYSTEM

The relationship between changes in the pressure distending the alveoli and changes in lung volume is important to understand because it dictates how the lung inflates with each breath. As mentioned before, the alveolar-distending pressure is often referred to as the *transpulmonary pressure*. Strictly speaking, the transpulmonary pressure is equal to the pressure in the trachea minus the intrapleural pressure. Thus, it is the pressure difference across the *whole lung*. The pressure in the alveoli, however, is the same as the pressure in the airways, including the trachea, at the beginning or end of each normal breath; that is, end-expiratory or end-inspiratory alveolar pressure is 0 cmH_2O, as can be seen in Fig. 2-5. Therefore, at the beginning or end of each lung inflation, alveolar-distending pressure can be referred to as the *transpulmonary pressure*.

Compliance of the Lung and Chest Wall

The pressure-volume characteristics of the lung can be inspected in several ways. One of the simplest is to remove the lungs from an animal and then graph the changes in volume that occur for each change in transpulmonary pressure the lungs are subjected to, as was done in Fig. 2-6.

Figure 2-6 shows that as the transpulmonary pressure increases, the lung volume increases. Of course, this relationship is not a straight line: The lung is composed of living tissue, and although the lung distends

Figure 2-6 Pressure-volume curve for isolated lungs.

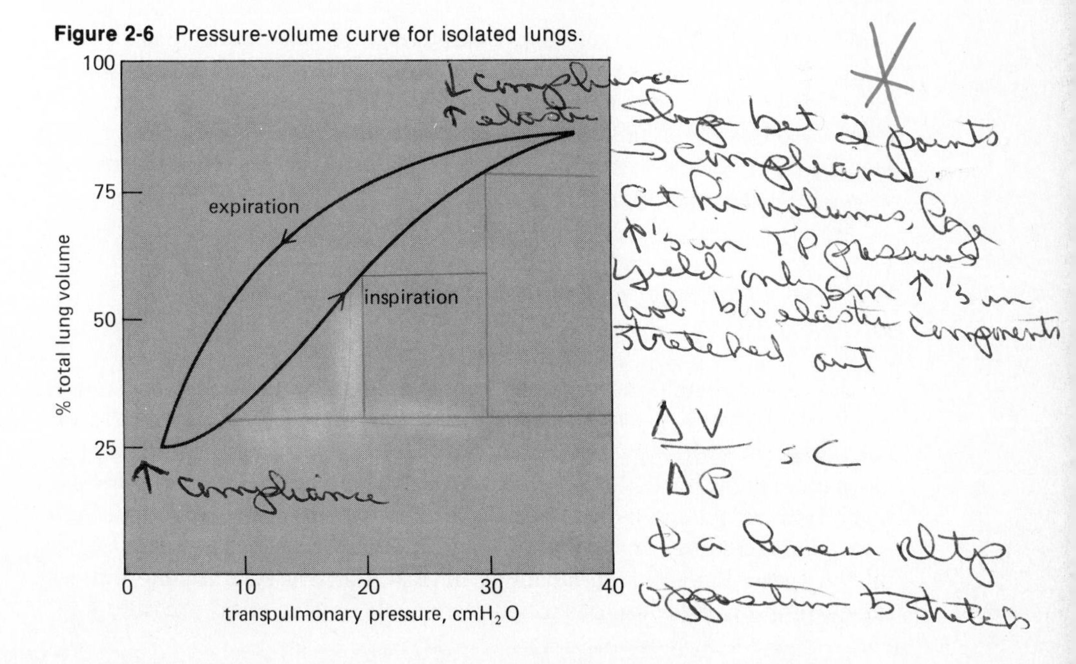

easily at low lung volumes, at high lung volumes the elastic components of alveolar walls have already been stretched out and large increases in transpulmonary pressure only yield small increases in volume.

The slope between two points on a pressure-volume curve is known as the *compliance*. Compliance is defined as the change in volume divided by the change in pressure. It is important to remember that compliance is the *inverse* of elasticity or elastic recoil. Compliance denotes the ease with which something can be stretched or distorted; elasticity refers to the tendency for something to oppose stretch or distortion, as well as to its ability to return to its original configuration after the distorting force is removed.

There are several other interesting things to note about an experiment like that illustrated in Fig. 2-6. The curve obtained is the same if the lungs are inflated with positive pressure (by forcing air into the trachea) or with negative pressure (by suspending the lung except for the trachea in a closed chamber and pumping out the air around the lung). So when the lung alone is considered, only the transpulmonary pressure is important, not how the transpulmonary pressure is generated. A second feature of the curve in Fig. 2-6 is that there is a difference between the pressure-volume curve for inflation and the curve for deflation, as shown by the arrows. Such a difference is called *hysteresis*. One possible explanation for this hysteresis will be given later in this chapter. Finally, it is helpful to think of each alveolus as having its own pressure-volume curve like that shown in Fig. 2-6.

Clinical Evaluation of the Compliance of the Lung and Chest Wall The compliance of the lung and chest wall provides very useful data for the clinical evaluation of a patient's respiratory system, because many diseases or pathological states affect either the compliance of the lung or chest wall or both. The lung and chest wall are physically in series with each other and therefore their compliances add as reciprocals:

$$\frac{1}{\text{Total compliance}} = \frac{1}{\text{compliance of the lung}} + \frac{1}{\text{compliance of the chest wall}}$$

To make clinical determinations of pulmonary compliance one must be able to measure changes in pressure and volume. Volume changes can be measured with a spirometer, but measuring the pressure changes is more difficult because changes in the *transmural pressure gradient* must be taken into account. For the lungs the transmural pressure gradient is the transpulmonary pressure (alveolar minus intrapleural); for the chest wall the transmural pressure gradient is intrapleural pressure minus atmospheric pressure. As was described previously, intrapleural pressure can be measured by having the patient swallow an esophageal balloon. The compliance curve for the lung can then be generated by having the patient take a very deep breath and then exhale in stages, stopping in midbreath for pressure and volume determinations. During these determinations, no airflow is occurring; alveolar pressure therefore equals atmospheric pressure, 0 cmH_2O. Such curves are called *static compliance* curves because all measurements are made when no airflow is occurring. The compliance of the chest wall is normally obtained by determining the compliance of the total system and the compliance of the lungs alone and then calculating the compliance of the chest wall according to the above formula. *Dynamic compliance,* in which pressure-volume characteristics *during* the breath are considered, will be discussed later in this chapter.

Representative static compliance curves for the lungs are shown in Fig. 2-7. Many pathological states shift the curve to the right (that is, for any increase in transpulmonary pressure there is less of an increase in lung volume). A proliferation of connective tissue, as is found in the fibrosis seen in sarcoidosis or seen after chemical or thermal injury to the lungs, will make the lungs less compliant or "stiffer." Similarly, pulmonary vascular engorgement or areas of collapsed alveoli (atelectasis) also makes the lung less compliant. Other conditions that interfere with the lung's ability to expand, such as air, excess fluid, or blood in the intrapleural space, will decrease the compliance of the lungs. Emphysema *in-*

Figure 2-7 Representative static pulmonary compliance curves for normal lungs, fibrotic lungs, and emphysematous lungs. (*Reproduced with permission from Murray, 1976.*)

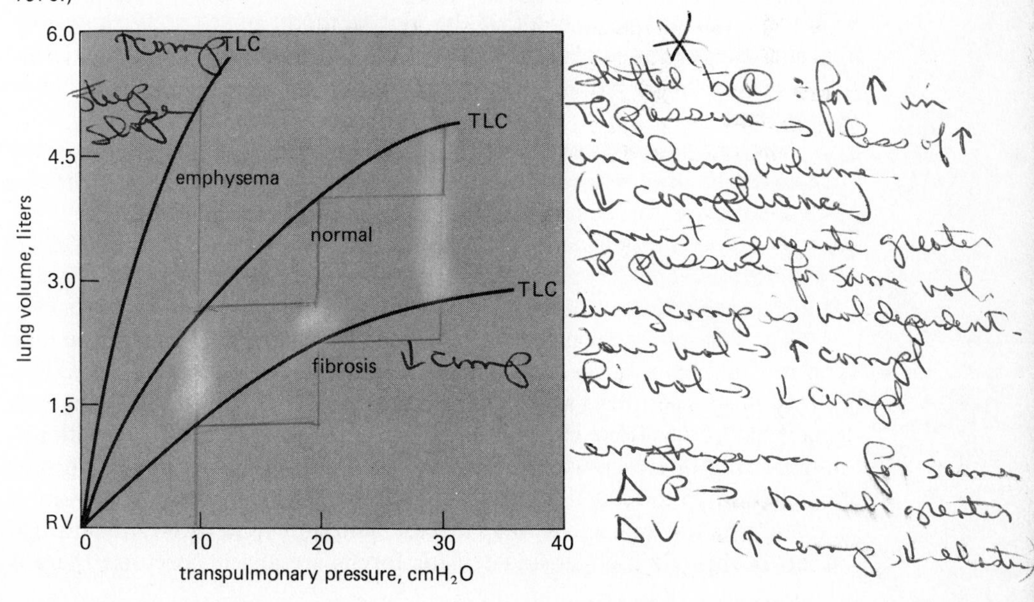

creases the compliance of the lungs because it destroys the alveolar septal tissue that normally opposes lung expansion.

The compliance of the chest wall is decreased in obese people, for whom moving the diaphragm downward and the rib cage up and out is much more difficult. People suffering from a musculoskeletal disorder that leads to decreased mobility of the rib cage, such as kyphoscoliosis, also have decreased chest wall compliance.

Because they must generate greater transpulmonary pressures to breathe in the same volume of air, persons with decreased compliance of the lungs must do more work to inspire than those with normal pulmonary compliance. Similarly, more muscular work must be done by someone with decreased chest wall compliance than by a person with normal chest wall compliance.

As noted in the beginning of this section, lung compliance is volume-dependent. It is greater at lower lung volumes and lower at high lung volumes. For this reason the term *specific compliance* is often used to denote compliance with reference to the original lung volume.

The total compliance of a normal person near the normal end-expiratory lung volume (the *functional residual capacity*) is about 0.1 liter/ cmH_2O. The compliance of the lungs is about 0.2 liter/cmH_2O; that of the chest wall is also about 0.2 liter/cmH_2O. (See Probs. 1 through 3 for more details on compliance measurement.)

Elastic Recoil of the Lung

So far we have discussed the elastic recoil of the lungs as though it was due only to the elastic properties of the pulmonary parenchyma itself. There is another component of the elastic recoil of the lung, however, besides the elastin, collagen, and other constituents of the lung tissue. That other component is the surface tension at the air-liquid interface in the alveoli.

Surface tension forces occur at any gas-liquid interface. They are generated by the cohesive forces between the molecules of the liquid. These cohesive forces balance each other within the liquid phase but are unopposed at the surface of the liquid. Surface tension is what causes water to bead and form droplets. It causes a liquid to shrink to form the smallest possible surface area. The unit of surface tension is dyn/cm.

The role of the surface tension forces in the elastic recoil of the lung can be demonstrated in an experiment such as that shown in Fig. 2-8.

In this experiment a pressure-volume curve for an excised lung was generated as was done in Fig. 2-6. Because the lung was inflated with air, an air-liquid interface was present in the lung and surface tension forces contributed to the elastic recoil of the lung. Then all of the gas was removed from the lung and it was inflated again, this time with saline instead of air. In this situation surface tension forces are absent, because there is

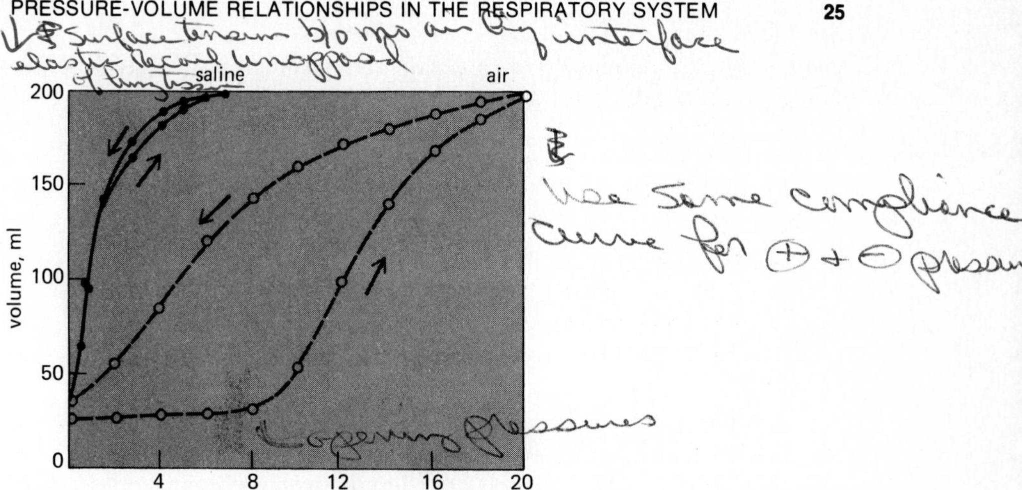

Figure 2-8 Pressure-volume curves for excised cat lungs inflated with air or saline. (*Reproduced with permission from Clements, 1965.*)

no air-liquid interface. The elastic recoil is due only to the elastic recoil of the lung tissue itself. Note that there is no hysteresis with saline inflation. Whatever causes the hysteresis appears to be related to surface tension in the lung. To recapitulate, the curve at left (saline inflation) represents the elastic recoil due to only the lung tissue itself. The curve at right demonstrates the recoil due to both the lung tissue and the surface tension forces. The difference between the two curves is the recoil due to surface tension forces.

The demonstration of the large role of surface tension forces in the recoil pressure of the lung led to consideration of how surface tension affects the alveoli. If the alveolus is considered to be a sphere, as in Fig. 2-9, then the relationship between the pressure inside the alveolus and the

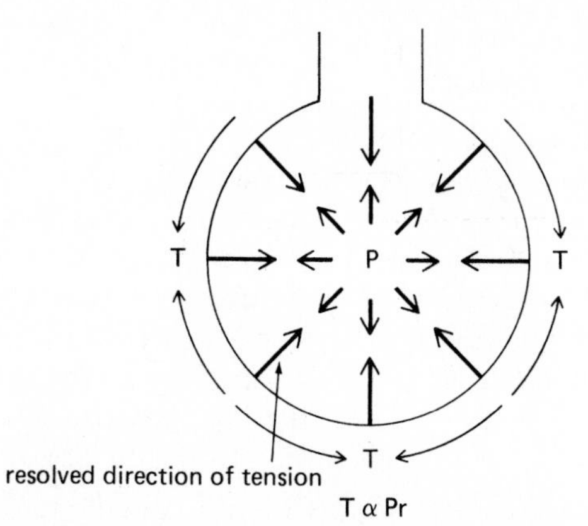

Figure 2-9 Relationship between the pressure inside a distensible sphere such as an alveolus and its wall tension.

wall tension of the alveolus is given by Laplace's law (units in parentheses):

$$\text{Pressure (dyn/cm}^2) = \frac{2 \times \text{tension(dyn/cm)}}{\text{radius (cm)}}$$

This can be rearranged as

$$T = \frac{Pr}{2}$$

The surface tension of most liquids (such as water) is constant and not dependent on the area of the air-liquid interface. Consider what this would mean in the lung, where alveoli of different sizes are connected to each other by common airways and collateral ventilation pathways, as described in Chap. 1. If we imagine that two alveoli of different sizes are connected by a common airway, as shown in Fig. 2-10, and that the surface tension of the two alveoli is equal, then according to Laplace's law the pressure in the smaller alveolus must be greater than that in the larger alveolus, and the smaller alveolus would empty into the larger alveolus. (The smaller the small alveolus is, the higher is the pressure in it.)

Thus, if the lung were composed of interconnected alveoli of different sizes (which it is) with a *constant surface tension* at the air-liquid interface, we would expect it to be inherently unstable, with a tendency for smaller alveoli to collapse into larger ones. Normally this is not the case, which is fortunate because collapsed alveoli require very great distending pressures to reopen, partly because of the cohesive forces at the liquid-liquid interface of collapsed alveoli. At least two factors cause the alveoli

Figure 2-10 Schematic representation of two alveoli of different sizes connected to a common airway. If the surface tension is the same in both alveoli, then the smaller alveolus will have a higher pressure and it will empty into the larger alveolus.

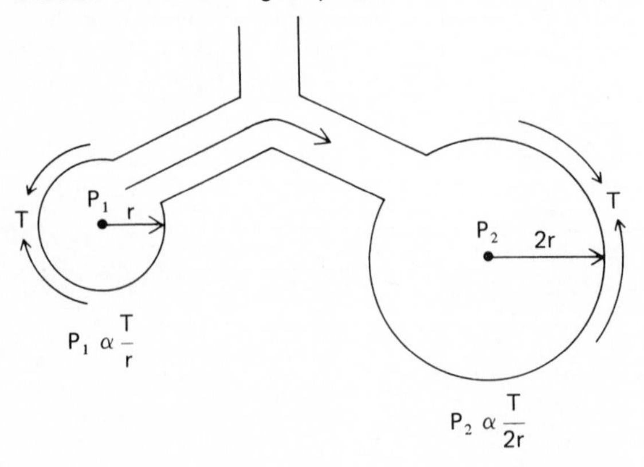

to be more stable than this prediction based on constant surface tension. The first of these is a substance known as *pulmonary surfactant,* which is produced by specialized alveolar cells; the second is the structural interdependence of the alveoli.

Pulmonary Surfactant

The surface tension of a liquid can be measured with an apparatus like that shown in Fig. 2-11.

The liquid to be inspected is placed in the trough. The movable barrier (denoted by the arrow at right) allows a determination of the role of the surface area of the air-liquid interface on surface tension. The surface tension is measured by the downward force on the platinum strip suspended from the force transducer, which is seen at left.

The results of a series of such experiments are shown in Fig. 2-12. The surface tension properties of water, water after the addition of detergent, blood serum, and lung extract are plotted with respect to the relative surface area of the trough seen in Fig. 2-11. Water has a relatively high surface tension, about 72 dyn/cm, and it is completely independent of surface area. Addition of detergent to the water decreases the surface tension, but it is still independent of surface area. Blood serum has a surface tension lower than that of water, and to a small extent its surface tension is dependent on surface area because its surface tension is somewhat lower at smaller relative areas. "Lung extract," which was obtained by washing the liquid film that lines the alveoli out with saline, displays both low overall surface tension and a great deal of area dependence. Its maximum surface tension is about 45 dyn/cm, which occurs at high relative areas. At low relative areas, the surface tension falls to nearly 0

Figure 2-11 A Langmuir-Wilhelmy balance for measurement of surface tension. (*Reproduced with permission from Clements, 1965.*)

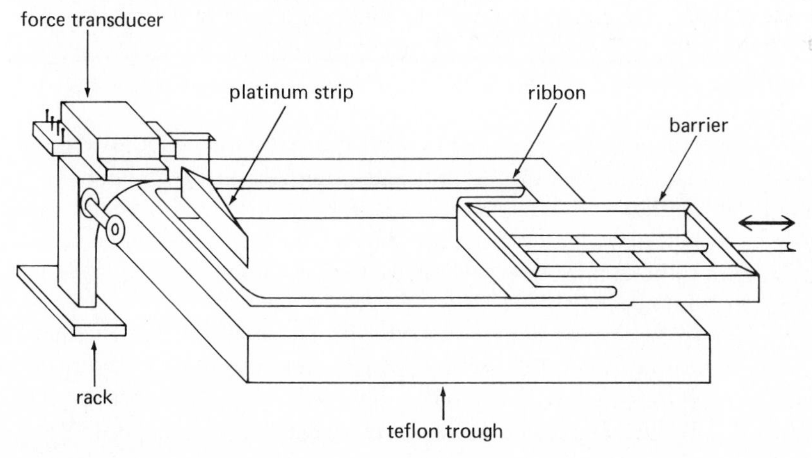

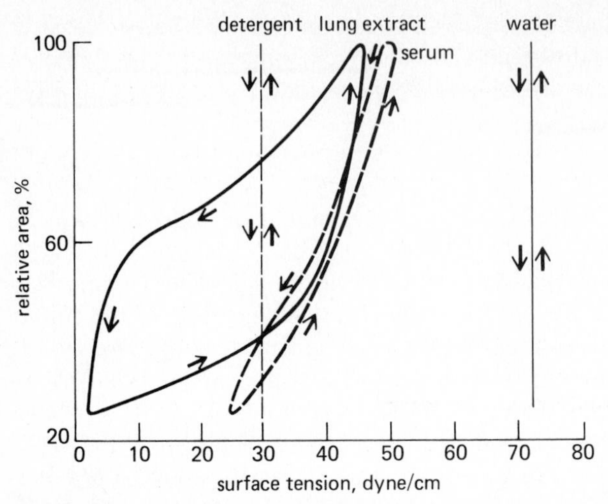

Figure 2-12 Schematic representation of surface tension–area relationships of water, water with detergent, serum, and lung extract. (*Reproduced with permission from Clements, 1965.*)

dyn/cm. Furthermore, the lung extract also displays a great deal of hysteresis, similar to that seen in Figs. 2-6 and 2-8.

From these data we can conclude that the alveoli are lined with a fluid that lowers their elastic recoil due to surface tension, even at high lung volumes. This increases the compliance of the lungs above that predicted by an air-water interface and thus decreases the work necessary to inspire. Because the surface tension decreased dramatically at low relative areas, it is reasonable to assume that the surface tension of different-sized alveoli is *not constant* and that smaller alveoli have lower surface tensions. This helps to equalize alveolar pressures throughout the lung (so the end-expiratory pressure of all the alveoli is 0 cmH_2O) and helps to stabilize alveoli. Finally, the hysteresis seen in lung pressure-volume curves like those in Fig. 2-6 appears to be a property of the fluid lining the alveoli, although the precise physicochemical reason for this is not yet fully understood.

The surface-active component of the lung extract is called *pulmonary surfactant.* It is a complex of different phospholipids consisting mainly of dipalmitoyl lecithin and it is manufactured by specialized alveolar cells known as type II alveolar epithelial cells (see Chap. 1). Pulmonary surfactant appears to be continually produced in and in some way destroyed or used up in the lung.

The clinical consequences of a lack of functional pulmonary surfactant can be seen in several conditions. Surfactant is not produced by the fetal lung until about the fourth month of gestation and it may not be fully functional until the seventh month or later. Prematurely born infants who do not have functional pulmonary surfactant experience great difficulty in inflating their lungs, especially on their first breath. Even if their alveoli are inflated for them, the tendency toward spontaneous collapse is great because their alveoli are much less stable without pulmonary surfactant. Therefore the lack of functional pulmonary surfactant may be a major factor in the *infant respiratory distress syndrome* of the prematurely born neonate.

Hypoxia, hypoxemia (low oxygen in the arterial blood), or both may lead to a decrease in surfactant production or an increase in surfactant destruction. This condition may be a contributing factor in the adult respiratory distress syndrome (sometimes called "shock-lung syndrome") seen in patients after trauma or surgery. One thing that can be done to help maintain such patients is to ventilate their lungs with positive-pressure ventilators and to keep their alveolar pressure above atmospheric pressure during expiration [this procedure is known as positive end-expiratory pressure (PEEP)]. This process opposes the increased elastic recoil of the alveoli and the tendency for spontaneous atelectasis to occur because of a lack of pulmonary surfactant.

In summary, pulmonary surfactant helps to lower the work of inspiration by lowering the surface tension of the alveoli, thus reducing the elastic recoil of the lung and making the lung more compliant. Surfactant also helps stabilize the alveoli by lowering even further the surface tension of smaller alveoli, equalizing the pressure inside alveoli of different sizes.

Alveolar Interdependence

A second factor tending to stabilize the alveoli is their mechanical interdependence, which was already discussed at the beginning of this chapter. If an alveolus, such as the one in the middle of Fig. 2-13, were to begin to collapse, it would increase the stresses on the walls of the alveoli adjacent to it, which would tend to hold it open. This process would oppose a tendency for isolated alveoli suffering from a relative lack of pulmonary surfactant to collapse spontaneously. Conversely, if a whole subdivision of the lung (such as a lobule) has collapsed, as soon as the first alveolus is

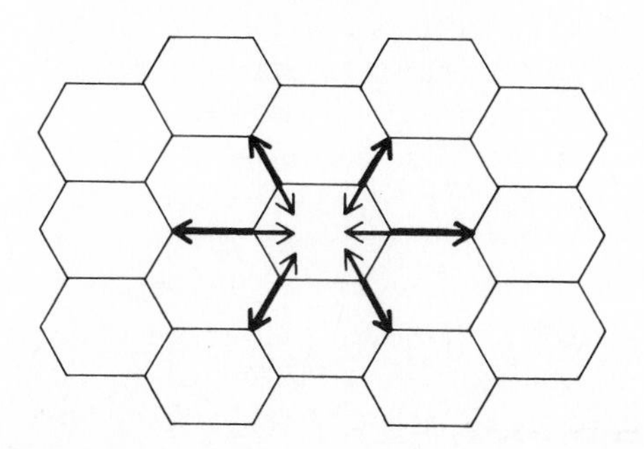

Figure 2-13. Representation of alveolar interdependence helping to prevent an alveolus from collapsing spontaneously.

reinflated it helps to pull other alveoli open by its mechanical interdependence with them.

INTERACTION OF LUNG AND CHEST WALL: THE STATIC PRESSURE-VOLUME CURVE

The interaction between the lung and chest wall was discussed earlier in this chapter. The inward elastic recoil of the lung opposes the outward elastic recoil of the chest wall, and vice versa. If the integrity of the lung–chest wall system is disturbed by opening the chest wall, as, for example, from a knife wound, the inward elastic recoil of the lung can no longer be opposed by the outward recoil of the chest wall and their interdependence ceases. Lung volume decreases, and alveoli have a much greater tendency to collapse, especially as air moves in through the wound until intrapleural pressure equalizes with atmospheric pressure and abolishes the transpulmonary pressure gradient. At this point nothing is tending to hold the alveoli open and their elastic recoil is causing them to collapse. Similarly, the chest wall tends to expand, because its outward recoil is no longer opposed by the inward recoil of the lung.

When the lung–chest wall system is intact and the respiratory muscles are relaxed, the volume of gas left in the lungs is *determined* by the balance of these two forces. The volume of gas in the lungs at the end of a normal expiration, when no respiratory muscles are contracting, is known as the *functional residual capacity* (FRC). For any given situation, the FRC will be that lung volume at which the outward recoil of the chest wall is equal to the inward recoil of the lungs. The relationship between lung elastic recoil and chest wall elastic recoil is illustrated in static (or "relaxation") pressure-volume curves such as those shown in Fig. 2-14.

Figure 2-14 Static volume-pressure curves of the lung, chest wall, and total system in the sitting and supine positions. (*Reproduced with permission from Agostoni, 1972.*)

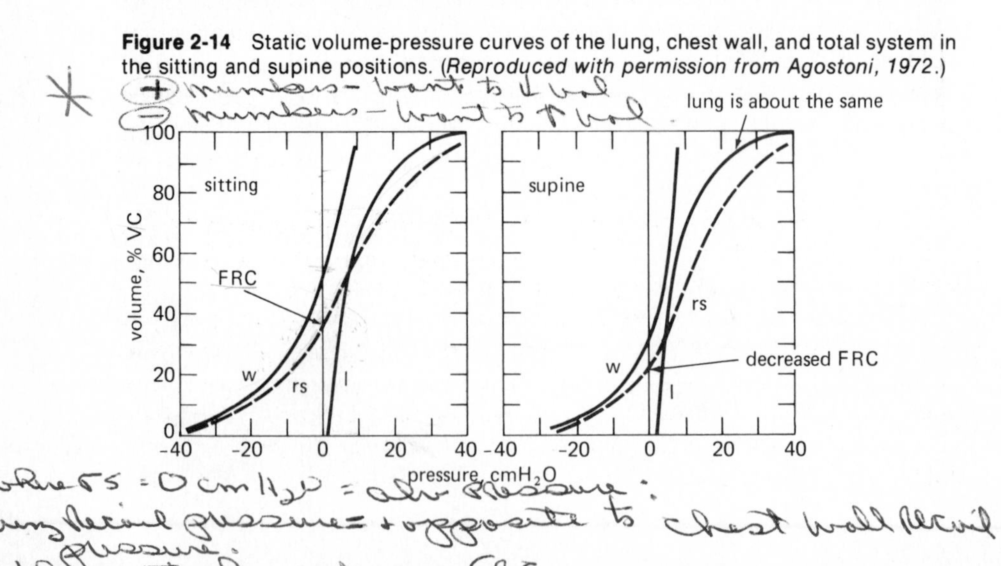

In the experiments from which these data were taken, subjects breathed air from a spirometer so that lung volumes could be measured. Intrapleural pressure was measured with an esophageal balloon and pressure was also measured at the subject's nose or mouth. The subjects were instructed to breathe air into or from the spirometer and then suddenly relax their respiratory muscles. The pressure then measured at the nose or mouth (which, at this point, when no airflow is occurring, is equal to alveolar pressure) is the sum of the recoil pressure of both lungs and the chest wall. It is represented by the dotted line labeled rs (respiratory system) in Fig. 2-14. The individual recoil pressures of the lung and chest wall can be calculated because the intrapleural pressure is known. Lung recoil pressure is labeled l on the graphs; chest wall recoil pressure is labelled w on the graphs. The graph on the left was drawn from data obtained when subjects were sitting up; the graph on the right was drawn from data obtained when subjects were lying on their backs.

The graph on the left shows that the pressure measured at the mouth (line rs) is equal to 0 cmH_2O at the point where lung recoil pressure is equal and opposite to chest wall recoil pressure. Therefore alveolar pressure is also 0 cmH_2O. The lung volume at this point is the subject's FRC.

As the subject increases his or her lung volume, the total system recoil pressure becomes positive because of two factors: the increased inward elastic recoil of the lung and the decreased outward elastic recoil of the chest wall. In fact, at high lung volumes the recoil pressure of the chest wall is also positive (note the point where the line w crosses the 0 pressure line). This is because at high lung volumes, above about 75 percent of the total lung capacity (TLC), the chest wall also has inward elastic recoil. In other words, if one could imagine a relaxed, intact chest wall with no lungs in it, the resting volume of the thorax would be about 75 percent of the volume of the thorax when the lungs are maximally (voluntarily) expanded. At thoracic volumes below about 75 percent of the total lung capacity, the chest wall elastic recoil is outward; at thoracic volumes above 75 percent of the TLC, the recoil is inward. Therefore at high lung volumes the mouth pressure is highly positive because both lung and chest wall elastic recoil is inward.

At lung volumes below the FRC the relaxation pressure measured at the mouth is negative because the outward recoil of the chest wall is now greater than the reduced inward recoil of the lungs.

The point of this discussion can be seen in the right-hand graph, in which the data collected were from supine subjects. Although the elastic recoil curve for the lung is relatively unchanged, the recoil curves for the chest wall and the respiratory system are shifted to the right. The reason for this shift is the effect of gravity on the mechanics of the chest wall, especially the diaphragm. When a person is standing up or sitting, the contents of the abdomen are being pulled away from the diaphragm by gravity. When the same person lies down, the abdominal contents are pushing inward against the relaxed diaphragm. This occurrence decreases the overall outward recoil of the chest wall and displaces the chest wall elastic recoil curve to the right. Because the respiratory system curve is the sum of the lung and chest wall curves, it is also shifted to the right.

The lung volume at which the outward recoil of the chest wall is equal to the inward recoil of the lung is much lower in the supine subject, as can be seen at the point where the rs crosses the 0 recoil pressure line. In other words, the functional residual capacity decreased appreciably just because the subject changed from the sitting to the supine position. Figure 2-15 shows the effect of body position on the FRC.

AIRWAYS RESISTANCE

Several factors besides the elastic recoil of the lungs and chest wall must be overcome to move air into or out of the lungs. These factors include the inertia of the respiratory system, the frictional resistance of the lung and chest wall tissue, and the frictional resistance of the airways to the flow of air. The inertia of the system is negligible. The lung-chest wall resistance is also normally quite negligible but may be greatly increased in patients suffering from mechanical limitations of the rib cage. Pulmonary tissue resistance is caused by the friction encountered as the lung tissues move against each other as the lung expands. The airways resistance plus the pulmonary tissue resistance is often referred to as the *pulmonary resis-*

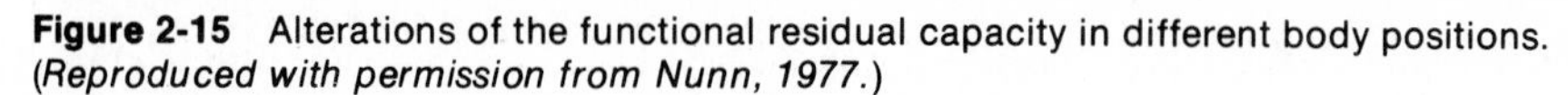

Figure 2-15 Alterations of the functional residual capacity in different body positions. (*Reproduced with permission from Nunn, 1977.*)

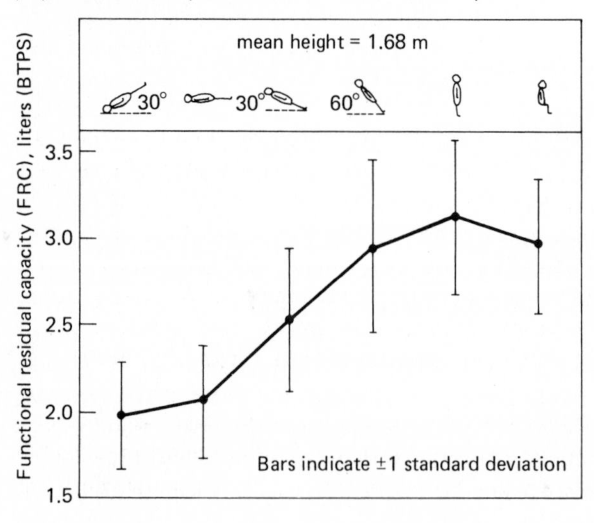

tance. Pulmonary tissue resistance normally contributes about 20 percent of the pulmonary resistance, with airways resistance responsible for the other 80 percent. Pulmonary tissue resistance can be increased in such conditions as pulmonary sarcoidosis or fibrosis. Because *airways resistance* is the major component of the total resistance and because it can increase tremendously both in normal people and in those suffering from various diseases, we will concentrate on airways resistance for the remainder of this chapter.

Laminar, Turbulent, and Transitional Flow

Generally, the relationship between pressure, flow, and resistance is stated as

$$\text{Pressure difference} = \text{flow} \times \text{resistance}$$

Therefore,

$$\text{Resistance} = \frac{\text{pressure difference (cmH}_2\text{O)}}{\text{flow (liters/s)}}$$

This means that resistance is a meaningful term only during flow. When airflow is considered, the units of resistance are usually cmH_2O/liter/s.

The resistance to airflow is analogous to electrical resistance in that resistances in *series* add as sums:

$$R_T = R_1 + R_2 + \cdots$$

Resistances in *parallel* add as reciprocals:

$$\frac{1}{R_T} = \frac{1}{R_1} + \frac{1}{R_2} + \cdots$$

Understanding and quantifying the resistance to airflow in the conducting system of the lungs is difficult because of the nature of the airways themselves. Although it is relatively easy to inspect the resistance to airflow in a single, unbranched, indistensible tube, the ever-branching, narrowing, distensible, and compressible system of airways makes analysis of the factors contributing to airways resistance especially complicated, and we will have to settle for approximations.

Airflow, like that of other fluids, can occur as either laminar flow or turbulent flow.

As seen in Fig. 2-16, *laminar* flow (or streamline flow) consists of a number of concentrically arranged cylinders of air flowing at different rates. This telescope-like arrangement is such that the cylinder closest to

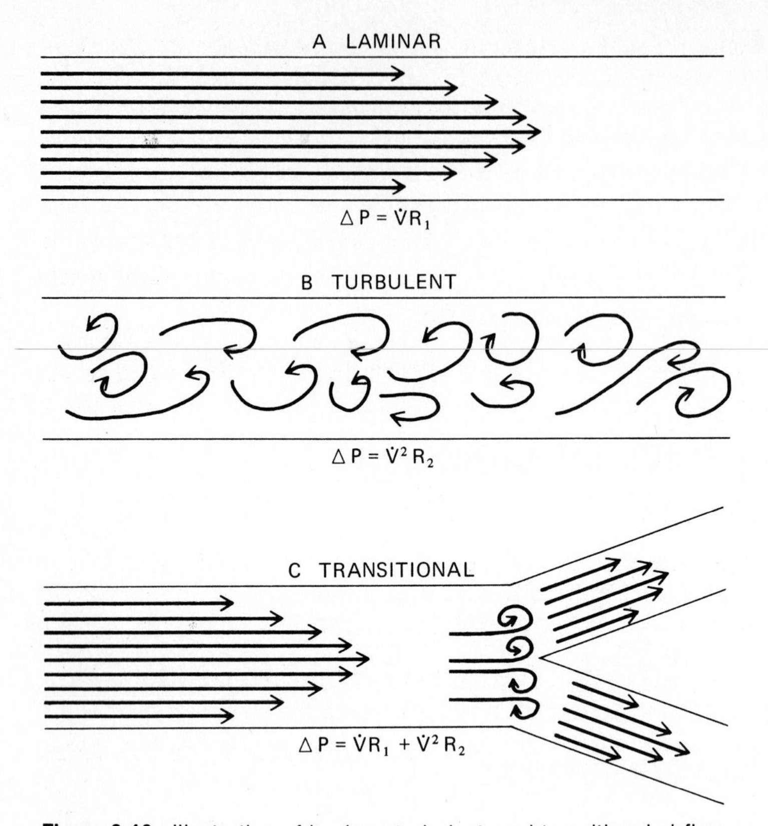

Figure 2-16 Illustration of laminar, turbulent, and transitional airflow.

the wall of the vessel has the slowest velocity because of frictional forces due to the wall; the pathway in the center of the vessel has the highest velocity.

When a fluid such as air flows through rigid, smooth-bore tubes, its behavior is governed by Poiseuille's law. The pressure difference is directly proportional to the flow times the resistance:

$$\Delta P = \dot{V}R_1$$

where ΔP = pressure difference
$\dot{V}$ = airflow
R_1 = resistance

According to Poiseuille's law, the resistance is directly proportional to the viscosity of the fluid and the length of the tube and is inversely proportional to the *fourth power* of the radius of the tube:

$$R = \frac{8\eta l}{\pi r^4}$$

where η = viscosity of the fluid
l = length of the tube
r = radius of the tube

Note that if the radius is cut in half, the resistance is multiplied by 16 because the resistance is inversely proportional to the radius to the fourth power.

Flow changes from *laminar* to *turbulent* when Reynold's number exceeds 2000. Reynold's number is a dimensionless number equal to the density of the fluid times the velocity of the fluid times the diameter of the tube divided by the viscosity of the fluid:

$$\text{Reynold's number} = \frac{\rho \times Ve \times D}{\eta}$$

where ρ = density of the fluid
Ve = linear velocity of the fluid
D = diameter of the tube
η = viscosity of the fluid

During turbulent flow the relationship between the pressure difference, flow, and resistance changes. The pressure difference is proportional to the flow *squared*. The resistance term is more influenced by the density than it is by the viscosity during turbulent flow:

$$\Delta P = \dot{V}^2 R_2$$

Transitional flow is a mixture of laminar and turbulent flow. This type of flow often occurs at branch points or points distal to partial obstructions.

Turbulent flow tends to occur if airflow is high, if gas density is high, if the tube radius is large, or if all three conditions exist. True laminar flow probably only occurs in the smallest airways, where the linear velocity of airflow is extremely low. Linear velocity (cm/s) is equal to the flow (cm^3/s) divided by the cross-sectional area. The total cross-sectional area of the smallest airways is very large (see Chap. 1), and so the linear velocity of airflow is very low. The airflow in the trachea and larger airways is usually either turbulent or transitional.

Distribution of Airways Resistance

About 25 to 40 percent of the total resistance to airflow is located in the upper airways: the nose, nasal turbinates, oropharynx, nasopharynx, and larynx. Resistance is higher when one breathes through the nose than when one breathes through the mouth.

As for the tracheobronchial tree, the component with the highest individual resistance is obviously the smallest airway, which has the smallest radius. Nevertheless, because the smallest airways are arranged in parallel, their resistances add as reciprocals, so that the total resistance to airflow offered by the numerous small airways is extremely low during normal, quiet breathing. Therefore, under normal circumstances the greatest resistance to airflow resides in the medium-sized bronchi.

Control of Bronchial Smooth Muscle

The smooth muscle of the airways from the trachea down to the alveolar ducts is under the control of efferent fibers of the autonomic nervous system, as was described in Chap. 1. Stimulation of the cholinergic *parasympathetic* postganglionic fibers causes *constriction* of bronchial smooth muscle, as well as increased glandular mucus secretion. The preganglionic fibers travel in the vagus. Stimulation of the adrenergic *sympathetic* fibers causes dilation of bronchial and bronchiolar smooth muscle, as well as inhibition of glandular secretion. This dilation of the airway smooth muscle is mediated by $beta_2$ receptors, which predominate in the airways. Selective stimulation of the alpha receptors with pharmacological agents causes bronchoconstriction.

Inhalation of chemical irritants, smoke, or dust, stimulation of the arterial chemoreceptors, and other stimulation, such as with histamine, cause *reflex constriction* of the airways. Decreased CO_2 in the branches of the conducting system causes a *local* constriction of the smooth muscle of the nearby airways. This may help to balance ventilation and perfusion when there is a pulmonary embolus (see Chap. 5).

Lung Volume and Airways Resistance

Airways resistance *decreases* with increasing lung volume, as shown in Fig. 2-17 (normal curve). This relationship is still present in an emphysematous lung, as will be discussed later in this chapter, although in emphysema the resistance is higher than that in the healthy state at all lung volumes.

There are two reasons for this relationship; both mainly involve the small airways, which, as described in Chap. 1, have little or no cartilaginous support. The small airways are therefore rather distensible (and also *compressible*). Thus the transmural pressure gradient across the wall of the small airways is an important determinant of the radius of the airways: Since resistance is inversely proportional to the radius to the *fourth power,* changes in the radii of small airways can cause dramatic changes in airways resistance, even with so many parallel pathways. To increase lung volume, a person breathing normally takes a "deep breath,"

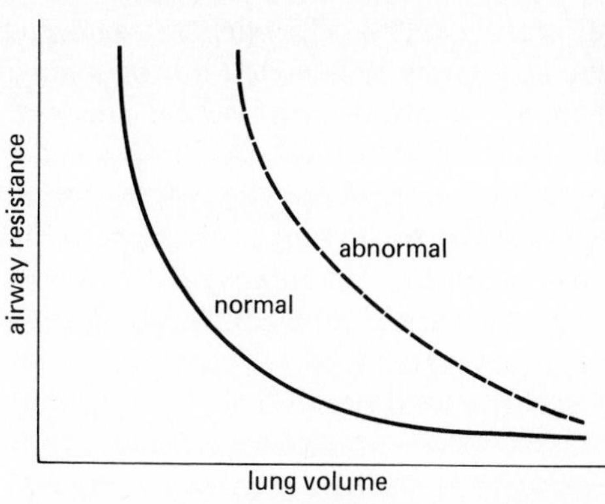

Figure 2-17 Relationship between lung volume and airways resistance. Total lung capacity is at right; residual volume is at left. Solid line = normal lung; dashed line = abnormal (emphysematous) lung. (*Reproduced with permission from Murray, 1972.*)

that is, he or she makes a strong inspiratory effort. This effort causes intrapleural pressure to become much more negative than the -7 or -10 cmH_2O seen in a normal, quiet breath. The transmural pressure gradient across the wall becomes much more positive and small airways are distended.

A second reason for the decreased airways resistance seen at higher lung volumes is that the so-called traction on the small airways increases. As shown in the schematic drawing in Fig. 2-18, the small airways traveling through the lung form attachments to the walls of alveoli. As the alveoli expand during the course of deep inspiration, the elastic recoil in their walls increases. This elastic recoil is transmitted to the attachments at the airway, pulling it open.

Dynamic Compression of Airways

Airways resistance is extremely high at low lung volumes, as can be seen in Fig. 2-17. To achieve low lung volumes, a person must make a forced expiratory effort by contracting the muscles of expiration, mainly the

Figure 2-18 Representation of "traction" of the alveolar septa on small distensible airways.

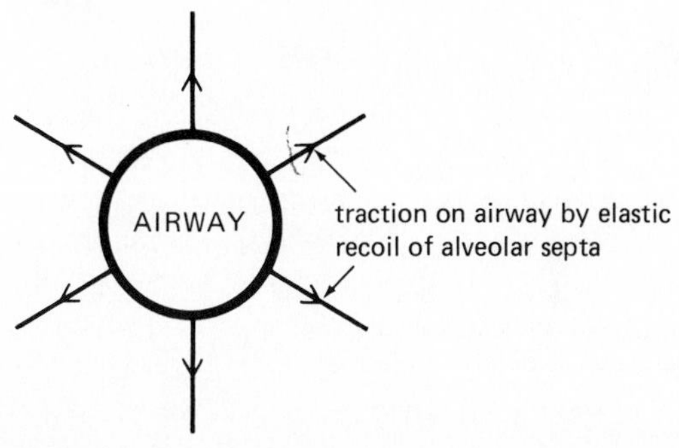

abdominal and internal intercostal muscles. This effort generates *positive* intrapleural pressure, which can be as high as 100 cmH_2O during a maximal forced expiratory effort. (Maximal inspiratory intrapleural pressures can be as low as -80 cmH_2O.)

The effect of this high positive intrapleural pressure on the transmural pressure gradient during a forced expiration can be seen at the right in Fig. 2-19, a schematic drawing of a single alveolus and airway.

The muscles of expiration are generating a positive intrapleural pressure of $+25$ cmH_2O. Pressure in the alveolus is higher than intrapleural pressure because of the alveolar elastic recoil pressure of $+10$ cmH_2O, which together with intrapleural pressure gives an alveolar pressure of $+35$ cmH_2O. The alveolar elastic recoil pressure decreases at lower lung volumes because the alveolus is not as distended. In the figure, a gradient has been established from the alveolar pressure of $+35$ cmH_2O to the atmospheric pressure of 0 cmH_2O. If the airways were rigid and incompressible, this large expiratory pressure gradient would generate very high rates of airflow. The airways are not, however, uniformly rigid and the smallest airways, which have no cartilaginous support and rely on the traction of

Figure 2-19 Schematic diagram illustrating dynamic compression of airways and the equal pressure point hypothesis during a forced expiration. *Left:* Passive (eupneic) expiration. *Right:* Forced expiration. Complete description in text.

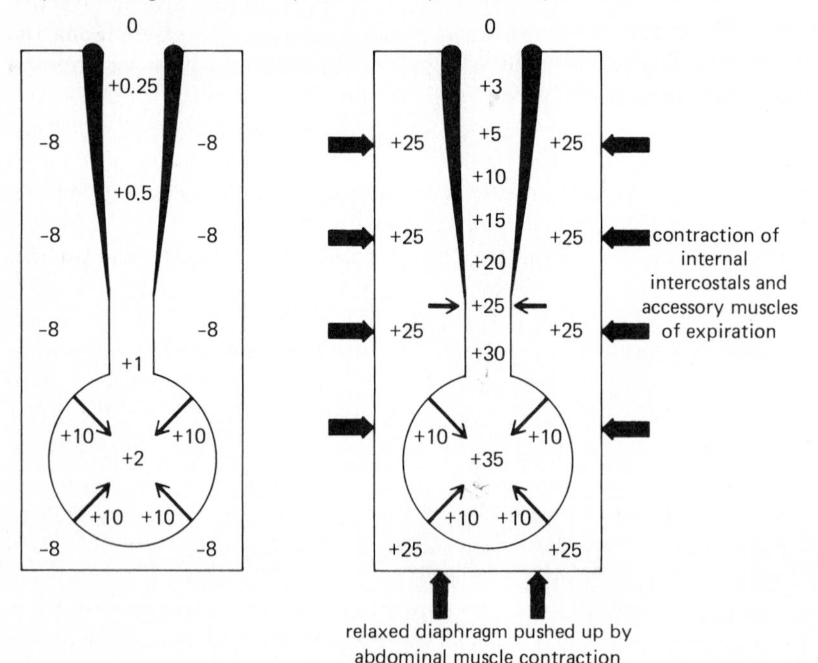

alveolar septa to help keep them open, may be compressed or even collapse. Whether or not they will collapse depends on the transmural pressure gradient across the walls of the smallest airways.

The situation during a normal *passive* expiration at the same lung volume (note the same alveolar recoil pressure) is shown in the left part of Fig. 2-19. The transmural pressure gradient across the smallest airway is

$$+1 \text{ cmH}_2\text{O} - (-8)\text{cmH}_2\text{O} = +9 \text{ cmH}_2\text{O}$$

tending to hold the airway open. During the *forced expiration* at right, the transmural pressure gradient is 30 cmH_2O − 25 cmH_2O or only 5 cmH_2O holding the airway open. The airway may then be slightly compressed and its resistance to airflow will be greater than during the passive expiration. This increased resistance during a forced expiration is called *dynamic compression* of airways.

Consider what must occur during a maximal forced expiration. As the expiratory effort is increased to attain a lower and lower lung volume, intrapleural pressure is getting more and more positive and more and more dynamic compression will occur. Furthermore, as lung volume decreases, there will be less alveolar elastic recoil pressure and the difference between alveolar pressure and intrapleural pressure will decrease.

One way of looking at this process is the *equal pressure point hypothesis*. At any instant during a forced expiration there is a point along the airways where the pressure inside the airway is just equal to the pressure outside the airway. At that point the transmural pressure gradient is 0 (note the arrows in Fig. 2-19). Above that point the transmural pressure gradient is *negative:* The pressure outside the airway is greater than the pressure inside it, and the airway will collapse if cartilaginous support or alveolar septal traction is insufficient to keep it open.

As the forced expiratory effort continues, the equal pressure point is likely to *move down the airway* from larger to smaller airways, as already stated. This movement happens because, as the muscular effort increases, intrapleural pressure increases and because, as lung volume decreases, alveolar elastic recoil pressure decreases. As the equal pressure point moves down the airway, dynamic compression increases and airways ultimately begin to collapse. This airway closure can be demonstrated only at especially low lung volumes in healthy subjects, but the *closing volume* may occur at higher lung volumes in patients with emphysema, as will be discussed at the end of this chapter. The closing volume test itself will be discussed in detail in Chap. 3.

It is important to consider the pressure gradient for airflow when thinking about dynamic compression and the equal pressure point hypothesis. During a passive expiration the pressure gradient for airflow (the ΔP in $\Delta P = \dot{V}R$) is simply alveolar pressure minus atmospheric pressure.

But if dynamic compression occurs, the effective pressure gradient is alveolar pressure minus *intrapleural pressure* because intrapleural pressure is higher than atmospheric pressure and because intrapleural pressure can exert its effects on the compressible portion of the airways.

Assessment of Airways Resistance

The resistance to airflow cannot be measured directly but must be calculated from the pressure gradient and airflow during a breath:

$$R = \frac{\Delta P}{\dot{V}}$$

This formula, of course, is an approximation because it presumes all air flow is laminar, which we know is not true. But there is a second problem: How can the pressure gradient be determined? To know the pressure gradient we must know alveolar pressure, which also cannot be measured directly. Alveolar pressure can be calculated using a body plethysmograph, an expensive piece of equipment described in detail in the next chapter, but this procedure is not often done. Instead, airways resistance is usually assessed indirectly. We will stress the assessment of airways resistance during expiration because that factor is of interest in patients with emphysema, chronic bronchitis, and asthma.

Forced Vital Capacity One way of assessing expiratory airways resistance is to look at the results of a forced expiration into a spirometer, as shown in Fig. 2-20. This measurement is called a *forced vital capacity* (FVC). The *vital capacity* is the volume of air a subject is able to expire after a maximal inspiration to the total lung capacity. A forced vital capacity means that a maximal expiratory effort was made during this maneuver.

After a few normal inspirations and expirations (seen at the far right of the upper curve in Fig. 2-20), the subject makes a maximal inspiration to the total lung capacity (TLC). After a moment, the subject makes a maximal forced expiratory effort, blowing as much air as possible out of the lungs. At this point only a *residual volume* (RV) of air is left in the lungs. (The lung volumes will be described in detail in the next chapter.) This procedure only takes a few seconds, as can be seen on the time scale.

The part of the curve most sensitive to changes in expiratory airways resistance is the first second of expiration. The volume of air expired in the first second of expiration (the FEV_1 or forced expiratory volume in 1 s), especially when expressed as a ratio with the total amount of air expired during the forced vital capacity, is a good index of expiratory airways resistance. In normal subjects, the FEV_1/FVC ratio is greater than 0.80; i.e., at least 80 percent of the forced vital capacity is expired in the first

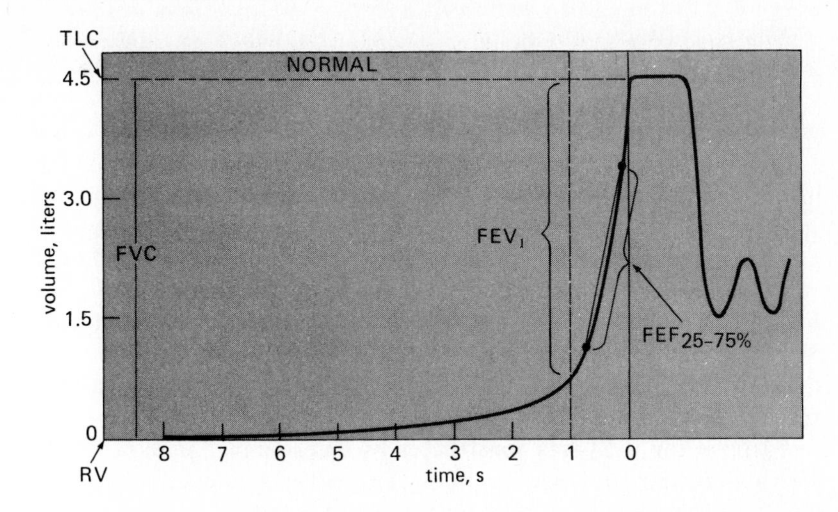

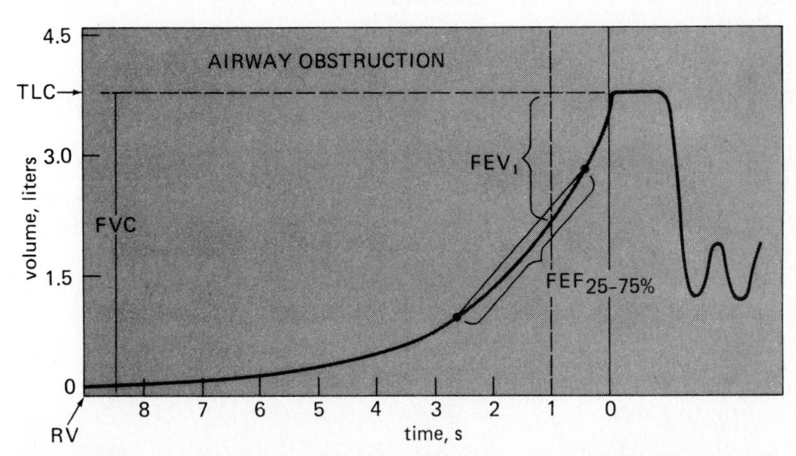

Figure 2-20 Forced vital capacity (FVC) maneuver using a spirometer. (See Fig. 3-4 for a diagram of a spirometer.) *Upper trace:* FVC from a normal subject. *Lower trace:* FVC from a patient with obstructive disease. FEV_1 = forced expiratory volume in the first second; $FEF_{25-75\%}$ = forced expiratory flow between 25 and 75% of the forced vital capacity.

second of the FVC. A patient with airway obstruction, caused, for example, by an episode of asthma, would be expected to have an FEV_1/FVC far below 0.80, as shown in the lower curve in Fig. 2-20.

Another way of expressing the *same* information is the $FEF_{25-75\%}$ or forced (mid) expiratory flow rate (formerly called the MMFR or maximal midexpiratory flow rate). This variable is simply the slope of a line drawn between the points on the expiratory curve at 25 and 75 percent of the

FVC. In cases of airway obstruction this line is not nearly as steep as it is on a curve obtained from someone with normal airways resistance.

Isovolumetric Pressure-Flow Curve This technique is not used often clinically because the data obtained are tedious to plot. Analysis of the results obtained from this test, however, demonstrates several points we have already discussed. Isovolumetric pressure-flow curves are obtained by having a subject make repeated expiratory maneuvers with different degrees of effort. Intrapleural pressures are determined with an esophageal balloon, lung volumes are determined with a spirometer, and airflow rates are determined by using a pneumotachograph. The pressure-flow relationship for each of the expiratory maneuvers of various efforts is plotted on a curve for a particular lung volume. For example, the middle curve of Fig. 2-21 was constructed by determining the intrapleural pressure and airflow for each expiratory maneuver as the subject's lung volume passed through 50 percent of the vital capacity.

The middle curve in Fig. 2-21 demonstrates dynamic compression and supports the equal pressure point hypothesis. At this lung volume, at which elastic recoil of the alveoli should be the same no matter what the

Figure 2-21 Isovolumetric pressure-flow curves at three different lung volumes: 75%, 50%, and 25% of the vital capacity. (*Reproduced with permission from Hyatt, 1965.*)

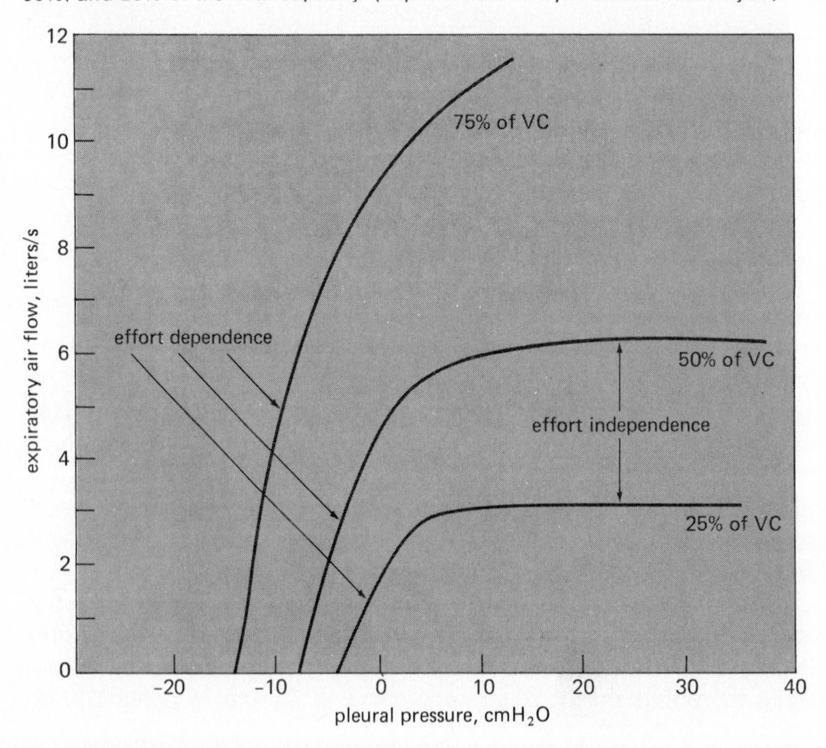

expiratory effort, with increasing expiratory effort airflow increases *up to a point*. Beyond that point, generating more positive intrapleural pressure does not increase airflow: It is *effort-independent*. Airways resistance must be increasing with increasing expiratory effort. Airflow has become independent of effort because of greater dynamic compression with more positive intrapleural pressures. The equal pressure point has moved to compressible small airways and is fixed there. Note that at even lower lung volumes (25 percent of VC), where there is less alveolar elastic recoil, this occurs with lower maximal airflow rates. At high lung volumes (75 percent VC), airflow increases steadily with increasing effort. It is entirely *effort-dependent* because alveolar elastic recoil pressure is high and because highly positive intrapleural pressures cannot be attained at such high lung volumes.

Flow-Volume Curves These same principles are demonstrated in the expiratory portion of flow-volume curves such as those in Fig. 2-22.

The family of flow-volume curves depicted in Fig. 2-22 was obtained in the same way as were the data in Fig. 2-21, only this time flow rates were plotted against lung volume for expiratory efforts of different intensities. Intrapleural pressures are not necessary. Because curves such as these can be plotted instantaneously if one has an XY plotter, this test is often used clinically. There are two interesting points about this family of curves, which correspond to the three curves in Fig. 2-21. At high lung volumes the airflow rate is effort-dependent, as can be seen in the left-hand portion of the curves. At low lung volumes, however, the expiratory efforts of different initial intensities all merge into the same effort-independent curve, as seen in the right-hand portion of the curve. Again, this difference is because high intrapleural pressures are necessary to attain very low lung volumes, no matter what the initial expiratory effort. Also, at low lung volumes there is less alveolar elastic recoil pressure. Both of these factors lead to dynamic compression of airways at low lung volumes.

The maximal flow-volume curve is often used as a diagnostic tool, as shown in Fig. 2-23, because it helps distinguish between two major classes of pulmonary diseases—airway *obstructive diseases* and *restrictive* diseases, such as fibrosis. Obstructive diseases are those diseases that interfere with airflow; restrictive diseases are those diseases that restrict the expansion of the lung.

Figure 2-23 shows that both obstruction and restriction can cause a decrease in the *maximal* flow rate that the patient can attain, but that this decrease occurs for different reasons. Restrictive diseases, which usually entail elevated alveolar elastic recoil, decrease the maximal flow rate because the total lung capacity (and thus the vital capacity) is decreased. The effort-independent part of the curve is similar to that obtained from a

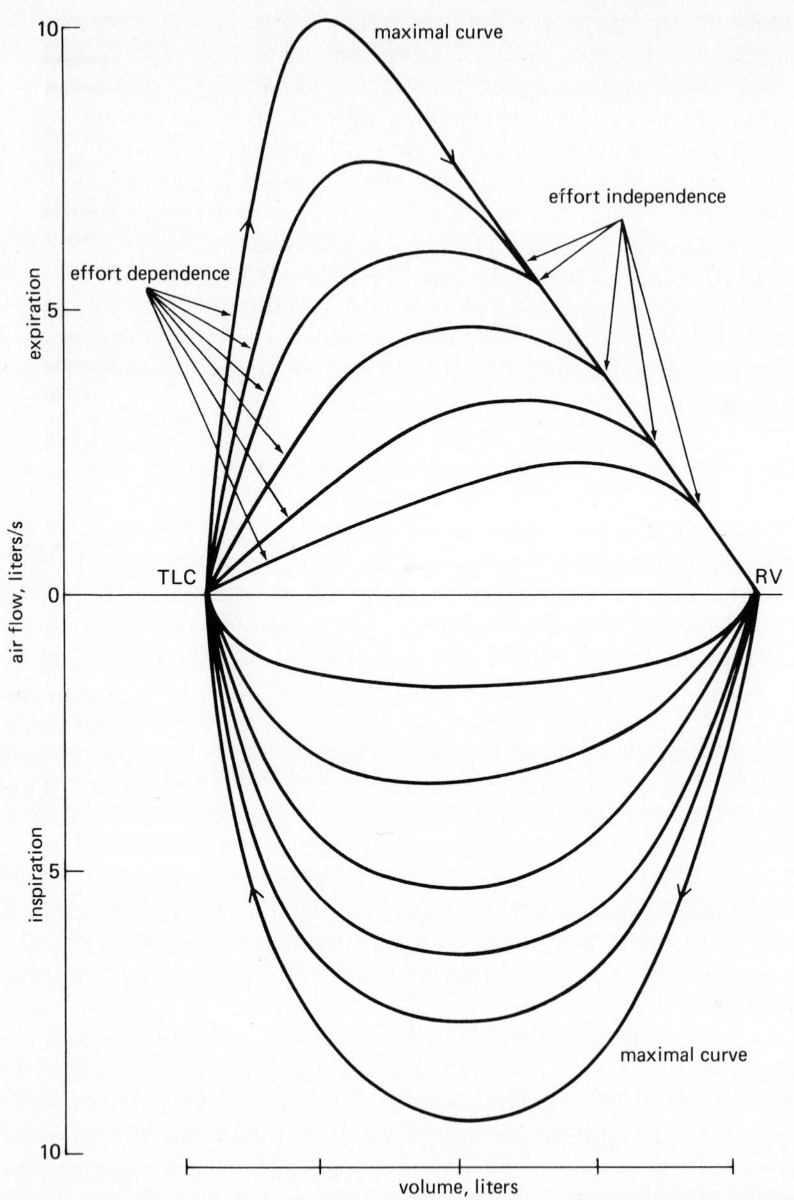

Figure 2-22 Flow-volume curves of varying intensities, demonstrating effort dependence at high lung volumes and effort independence at low lung volumes. Note that there is no effort independence in inspiration.

person with normal lungs. In fact, the FEV_1/FVC is usually normal or even above normal since *both* the FEV_1 and FVC are decreased because the lung has a low volume and because alveolar elastic recoil pressure may be increased. Obstructive diseases, such as asthma, bronchitis, and

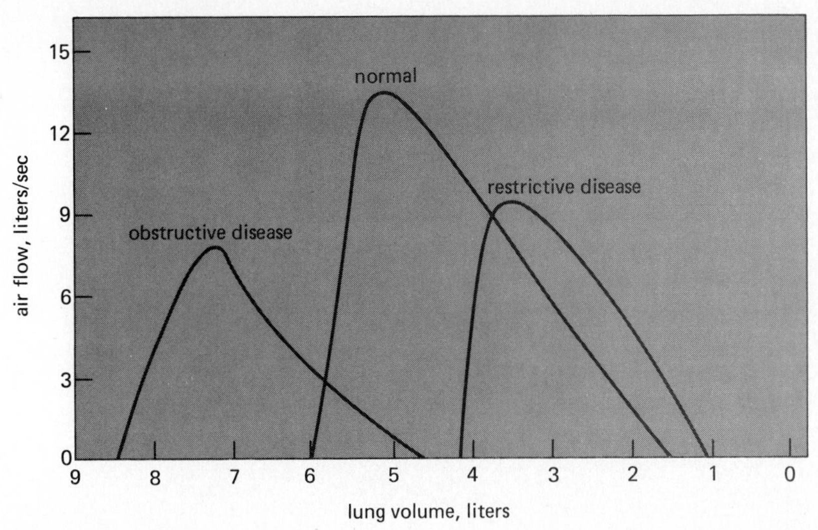

Figure 2-23 Maximal expiratory flow-volume curves representative of obstructive and restrictive diseases.

emphysema, are often associated with high lung volumes, which is helpful because the high volumes increase the alveolar elastic recoil pressure. The residual volume may be greatly increased if airway closure occurs at relatively high lung volumes. A second important feature of the flow-volume curve of a patient with obstructive disease is the effort-independent portion of the curve, which is depressed inward: Flow rates are *low* for any relative volume.

Dynamic Compliance

At this point we can consider the *dynamic compliance* of the lungs, which is the change in the volume of the lungs divided by the change in the alveolar-distending pressure *during the course of a breath*. At low breathing frequencies, around 15 breaths per minute and lower, dynamic compliance is about equal to static compliance and the ratio of dynamic compliance to static compliance is 1, as seen in Fig. 2-24.

In normal subjects this ratio stays near 1, even at much higher breathing frequencies. In patients with elevated resistance to airflow in some of their small airways, however, the ratio of dynamic compliance to static compliance falls dramatically as breathing frequency is increased. This occurrence indicates that changes in dynamic compliance reflect changes in airways resistance as well as changes in the compliance of alveoli.

The explanation for these changes is seen in Fig. 2-25. At the right in each figure is a pair of hypothetical alveoli supplied by the same airway. Consider the time courses of their changes in volume in response to an abrupt increase in airway pressure (a "step" increase) in a situation

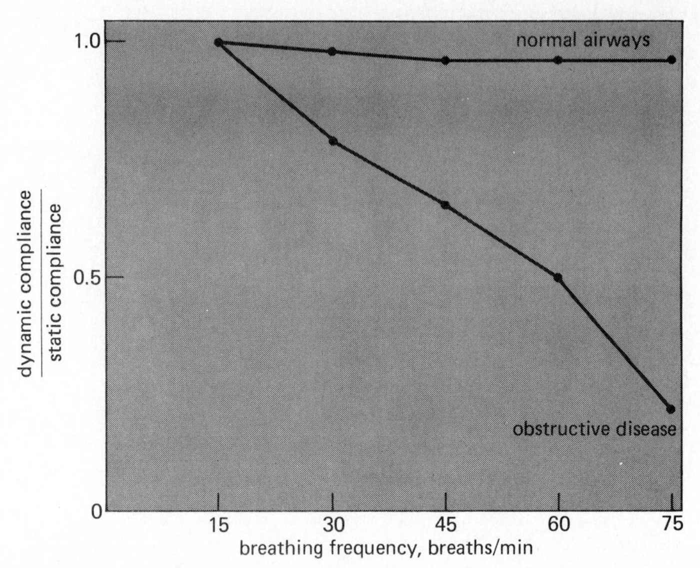

Figure 2-24 Schematic representation of changes in the ratio of dynamic compliance to static compliance with increasing breathing frequencies. The ratio changes little in normal subjects but decreases dramatically in patients with obstructive diseases of the small airways.

where we can arbitrarily alter the compliance of each alveolus or the resistance in the branch of the airway supplying it. The changes in volume with respect to time are shown at left.

In Fig. 2-25*A* the resistances and compliances of the two units are equal, and so the two units fill with identical time courses. In Fig. 2-25*B* the resistances are equal, but the compliance of *b* is one-half that of *a*. They fill with nearly identical time courses (*b* fills a little faster), but *b* receives only one-half the volume received by *a*. In Fig. 2-25*C* the compliances of the two units are equal, but *b* is supplied by an airway with twice the resistance to airflow of the one supplying *a*. The two units ultimately fill to the same volume (because they have equal compliances), but *b* fills more slowly than *a* because of its elevated resistance. In this hypothetical situation it took *b* 4 s to fill to the same final volume reached by *a* in a little more than 2 s. This difference means that at high breathing frequencies *a*, which fills much more quickly than *b*, will accommodate a larger volume of air per breath.

This situation may also lead to a redistribution of alveolar air after the inflating pressure has ceased. Consider the situation depicted in Fig. 2-25*C* if the inflating pressure is stopped at 2 s. *a* has more air in it than *b*, but both have equal compliance characteristics. *a* is more distended and therefore has a higher elastic recoil pressure than *b*. Because they are

A EQUAL COMPLIANCE, EQUAL RESISTANCE

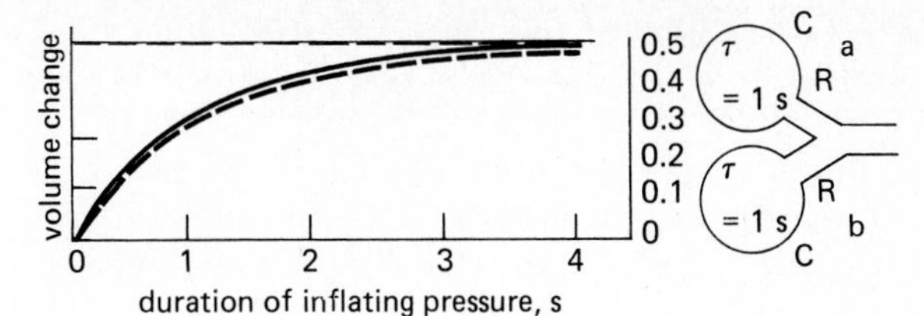

B EQUAL RESISTANCE, UNIT b HALF AS COMPLIANT AS a

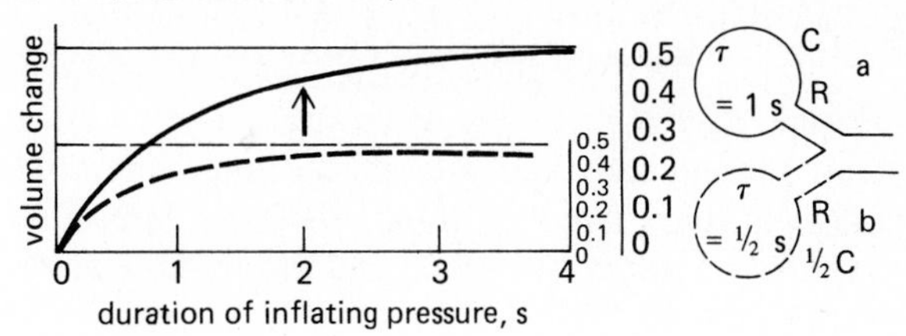

C EQUAL COMPLIANCE, UNIT b TWICE AS RESISTANT AS a

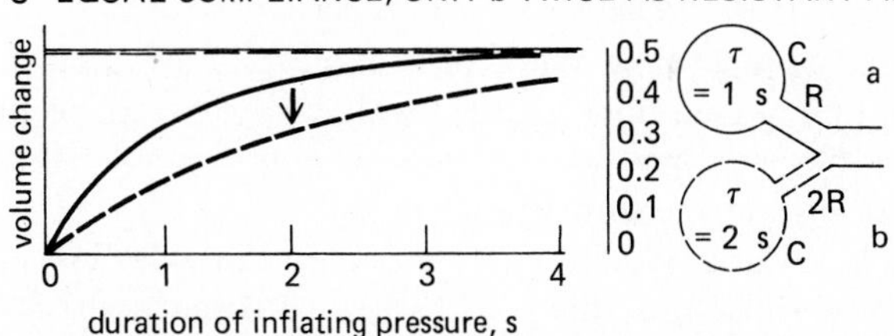

Figure 2-25 Time courses of volume changes for hypothetical alveoli of differing compliance supplied by airways of differing resistance. (*Adapted from Nunn, 1977, with permission.*)

joined by a common airway, some air is likely to follow the pressure gradient and move from *a* to *b*, as denoted by the arrow.

Now let's extrapolate this two-unit situation to a lung with millions of airways supplying millions of alveoli. In a patient with small-airways disease many alveoli may be supplied by airways with higher resistance to airflow than normal. These alveoli are sometimes referred to as "slow alveoli" or alveoli with long "time constants." (The time constant τ is equal to the product of the resistance and compliance and is about equal to the time required for inflation to 63 percent of the final volume attained if the inflating pressure is prolonged indefinitely. The time constants for the hypothetical alveoli in Fig. 2-25 are shown inside the alveoli.) As the patient increases the breathing frequency, the slowest alveoli will not *have enough time* to fill and will contribute nothing to the dynamic compliance. As the frequency increases, more and more slow alveoli will drop out and dynamic compliance continues to fall. Alveoli with low compliance can also affect the dynamic compliance because they do fill a little faster, as we saw in Fig. 2-25*B*, but their low static compliance prevents their contributing much to the dynamic compliance even at low breathing frequencies.

THE WORK OF BREATHING

We can now summarize the major points discussed in this chapter by considering the work of breathing. The work done in breathing is proportional to the pressure change times the volume change. The volume change is the volume of air moved into and out of the lung—the *tidal volume*. The pressure change is the change in transpulmonary pressure necessary to overcome the *elastic* work of breathing and the *resistive* work of breathing.

Elastic Work

The elastic work of breathing is the work done to overcome the elastic recoil of the chest wall and the pulmonary parenchyma and to overcome the surface tension of the alveoli. The work of breathing is elevated in obese patients, who have increased inward chest wall elastic recoil, or in patients with pulmonary fibrosis or a relative lack of pulmonary surfactant, who have increased elastic recoil of the alveoli.

The elastic work of breathing is the area bounded by line ABCDA in Fig. 2-26; the increased work of breathing in restrictive diseases is depicted in the middle panel.

Resistive Work

The resistive work of breathing is the work done to overcome the tissue resistance and airways resistance. The tissue resistance may be elevated in conditions such as sarcoidosis. Elevated airways resistance is much more common and is seen in asthma, bronchitis, and emphysema, as well as in accidental aspirations of foreign objects. Normally, most of the resistive work is that done to overcome airways resistance. The *additional* work of breathing necessary to overcome airways (and tissue) resistance on inspiration is the area bounded by line AA′BA. The area bounded by ABB′A is the work done on expiration to overcome airway (and tissue) resistance. Because this area falls inside the elastic work curve ABCDA, the expiration can be accomplished by the potential energy stored in the distended alveoli. The elevated work of breathing in obstructive disease is shown in the bottom panel.

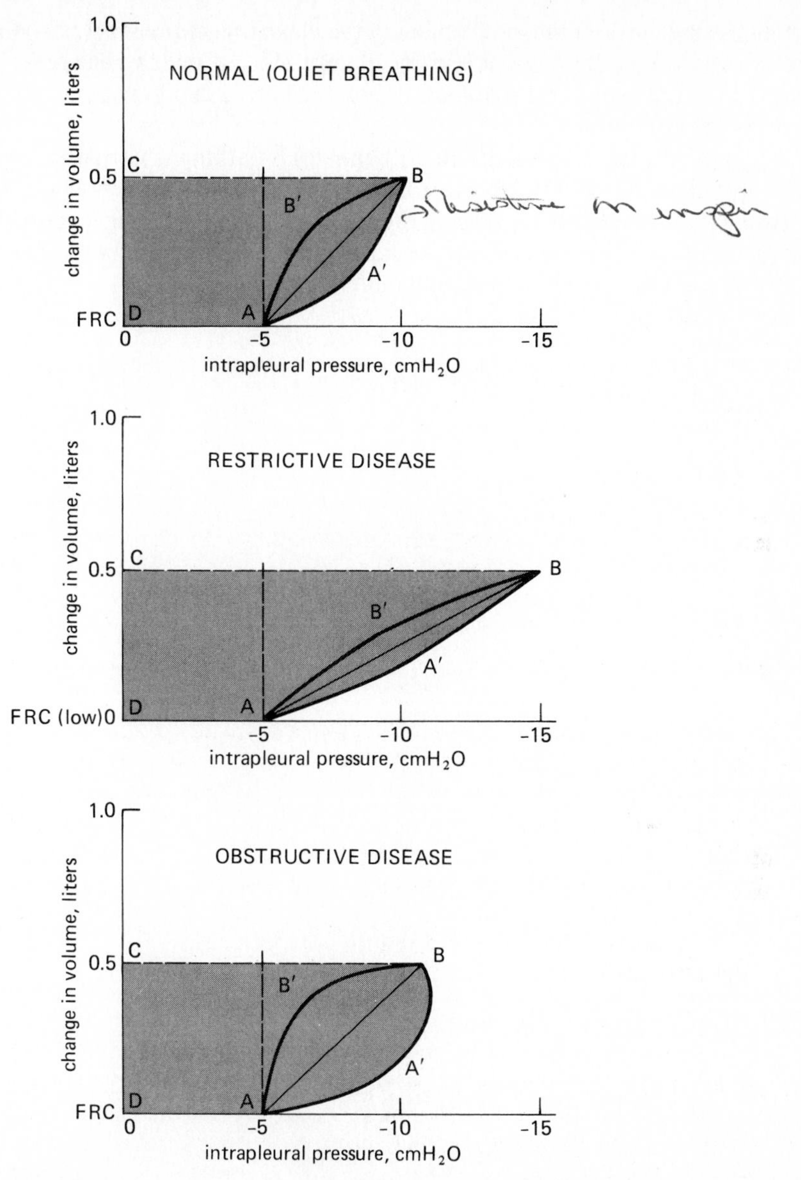

Figure 2-26 Schematic representation of the work of breathing in normal subjects, subjects with restrictive diseases, and subjects with obstructive diseases.

The resistive work of breathing can be extremely great during a *forced expiration,* when dynamic compression occurs. This is especially true in patients who already have elevated airways resistance during normal, quiet breathing. For example, in patients with emphysema, a disease that attacks and obliterates alveolar walls, the work of breathing can be tremendous because of the destruction of the elastic tissue support of their small airways, which allows dynamic compression to occur unopposed. Also, the decreased elastic recoil of alveoli leads to a decreased pressure gradient for expiration.

The oxygen cost of normal, quiet (eupneic) breathing is normally less than 5 percent of the total body oxygen uptake. This percentage can increase to as much as 30 percent in normal subjects during maximal exercise. In patients with obstructive lung disease, however, the work of breathing can be the factor that limits exercise.

Chapter 3

Alveolar Ventilation

OBJECTIVES

The student understands the ventilation of the alveoli.

1. Defines alveolar ventilation.
2. Defines the standard lung volumes and understands their measurement.
3. Predicts the effects of alterations in lung and chest wall mechanics, due to normal or pathological processes, on the lung volumes.
4. Defines anatomic dead space and relates the anatomic dead space and the tidal volume to alveolar ventilation.
5. Understands the measurement of the anatomic dead space and the determination of alveolar ventilation.
6. Defines physiological and alveolar dead space and understands their determination.
7. Predicts the effects of alterations of alveolar ventilation on alveolar carbon dioxide and oxygen levels.
8. Describes the regional differences in alveolar ventilation found in the normal lung and explains these differences.
9. Predicts the effects of changes in lung volume, aging, and disease processes on the regional distribution of alveolar ventilation.

10 Defines the closing volume and explains how it can be demonstrated.
11 Predicts the effects of changes in pulmonary mechanics on the closing volume.

Alveolar ventilation is the exchange of gas between the alveoli and the external environment. It is the process by which oxygen is brought into the lung from the atmosphere and by which the carbon dioxide carried into the lungs in the mixed venous blood is expelled from the body. Although alveolar ventilation is usually defined as the volume of fresh air *entering* the alveoli per minute, a similar volume of alveolar air *leaving* the body per minute is implicit in this definition.

THE LUNG VOLUMES

The volume of gas in the lungs at any instant depends on the mechanics of the lungs and chest wall and the activity of the muscles of inspiration and expiration. The lung volume under any specified set of conditions can be altered by pathological and normal physiological processes. Standardization of the conditions under which lung volumes are measured allows comparisons to be made between subjects or patients. Because the size of a person's lungs depends on the height and weight or body surface area of the person, as well as on his or her age and sex, the lung volumes for a patient are usually compared with data on a table of "predicted" lung volumes matched to age, sex, and body size. The lung volumes are normally expressed at body temperature and ambient pressure and water vapor saturation (BTPS).

The Standard Lung Volumes and Capacities

There are four standard lung *volumes,* which are not further subdivided, and four standard lung capacities, which consist of two or more standard lung volumes in combination, as shown in Fig. 3-1.

The Tidal Volume The tidal volume (TV or V_T) is the volume of air entering or leaving the nose or mouth per breath. It is determined by the activity of the respiratory control centers in the brain as they affect the respiratory muscles and by the mechanics of the lung and chest wall. During normal, quiet breathing (eupnea) the tidal volume of a 70-kg adult is about 500 ml per breath, but this volume can be greatly increased, for example, during exercise.

The Residual Volume The residual volume (RV) is the volume of gas left in the lungs after a maximal forced expiration. It is determined by the force generated by the muscles of expiration and the inward elastic recoil of the lungs as they oppose the outward elastic recoil of the chest wall.

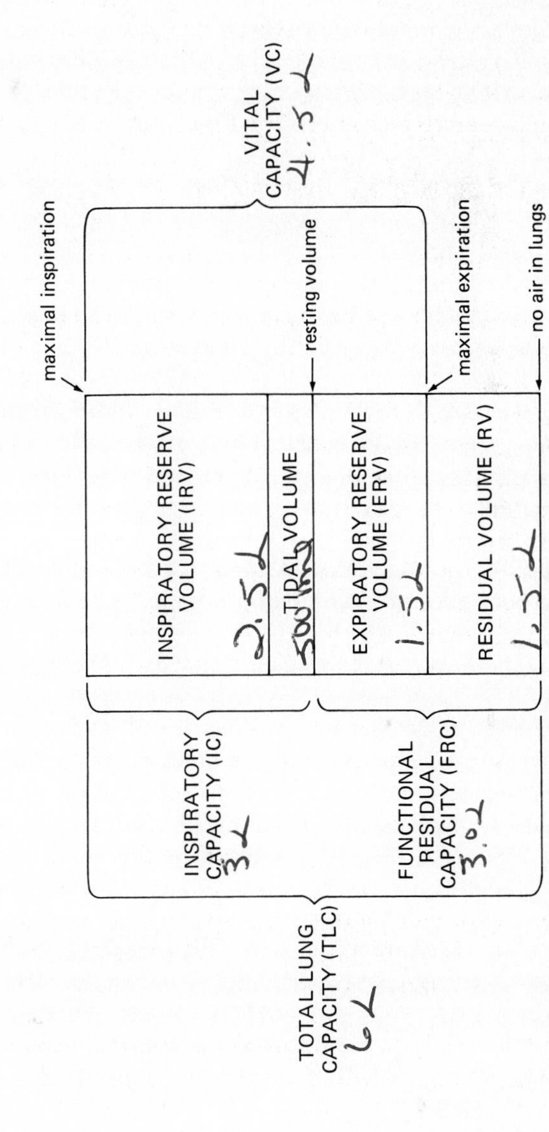

Figure 3-1 The standard lung volumes and capacities.

Dynamic compression of the airways during the forced expiratory effort may also be an important determinant of the residual volume as airway collapse occurs, thus trapping gas in the alveoli. The residual volume of a healthy 70-kg adult is about 1.5 liters, but it can be much greater in a disease state such as emphysema, in which inward alveolar elastic recoil is diminished and much airway collapse and gas-trapping occur. The residual volume is important to a healthy person because it prevents the lungs from collapsing at very low lung volumes. Such collapsed alveoli would require extremely great inspiratory efforts to reinflate.

The Expiratory Reserve Volume The expiratory reserve volume (ERV) is the volume of gas that is expelled from the lungs during a maximal forced expiration, which *starts* at the end of a normal tidal expiration. It is therefore determined by the difference between the functional residual capacity and the residual volume. The expiratory reserve volume is about 1.5 liters in a healthy 70-kg adult.

The Inspiratory Reserve Volume The inspiratory reserve volume (IRV) is the volume of gas that is inhaled into the lungs during a maximal forced inspiration *starting* at the end of a normal tidal inspiration. It is determined by the strength of contraction of the inspiratory muscles, the inward elastic recoil of the lung and chest wall, and the starting point, which is the functional residual capacity plus the tidal volume. The inspiratory reserve volume of a normal 70-kg adult is about 2.5 liters.

The Functional Residual Capacity The functional residual capacity (FRC) is the volume of gas remaining in the lungs at the end of a normal tidal expiration. Because no muscles of respiration are contracting at this time, it represents the balance point between the inward elastic recoil of the lungs and the outward elastic recoil of the chest wall, as discussed in Chap. 2. The FRC, as seen in Fig. 3-1, consists of the residual volume plus the expiratory reserve volume. It is therefore about 3 liters in a healthy 70-kg adult.

The Inspiratory Capacity The inspiratory capacity (IC) is the volume of air that is inhaled into the lungs during a maximal inspiratory effort, which *begins* at the end of a normal tidal expiration (the FRC). It is therefore equal to the tidal volume plus the inspiratory reserve volume, as shown in Fig. 3-1. The inspiratory capacity of a normal 70-kg adult is about 3 liters.

The Total Lung Capacity The total lung capacity (TLC) is the volume of air in the lungs after a maximal inspiratory effort. It is determined by the strength of contraction of the inspiratory muscles and the inward elastic recoil of the lungs and chest wall. The total lung capacity consists of all four lung volumes: the residual volume, the tidal volume, and the inspiratory and expiratory reserve volumes. It is about 6 liters in the healthy 70-kg adult.

The Vital Capacity The vital capacity (VC), already discussed in Chap. 2, is the volume of air expelled from the lungs during a maximal forced expiration starting after a maximal forced inspiration. The vital capacity is therefore equal to the total lung capacity minus the residual volume, or about 4.5 liters in a healthy 70-kg adult. The vital capacity is also equal to the sum of the tidal volume and the inspiratory and expiratory reserve volumes. It is determined by the factors that determine the total lung capacity and the residual volume.

MEASUREMENT OF THE LUNG VOLUMES

Measurement of the lung volumes is important clinically because many pathological states can alter specific lung volumes or their relationships to each other. The lung volumes, however, can also change for normal physiological reasons. Changing from a standing to a supine posture decreases the functional residual capacity because gravity is no longer pulling the abdominal contents away from the diaphragm. This decreases the outward elastic recoil of the chest wall, as noted in Chap. 2, Fig. 2-14. If the functional residual capacity is decreased, then the expiratory reserve volume will also decrease, as shown in Fig. 3-2, and the inspiratory reserve volume will increase. The vital capacity may decrease slightly because

Figure 3-2 Illustration of alterations in the lung volumes and capacities in changing from the standing to the supine position.

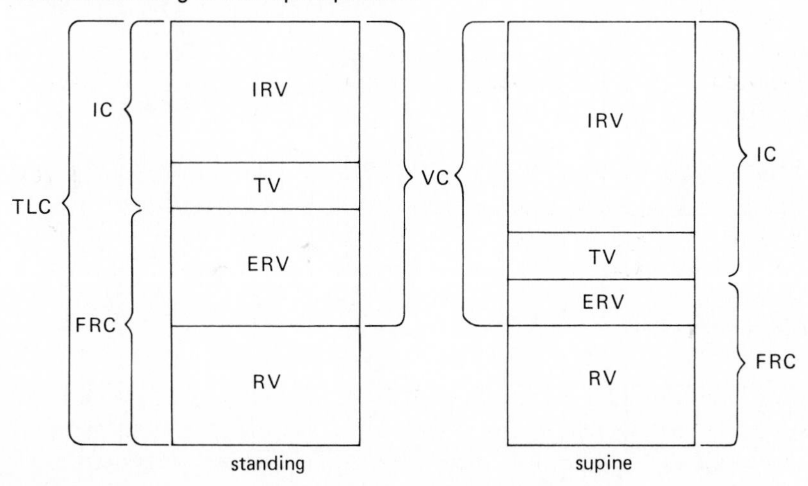

some of the venous blood that collects in the lower extremities and the abdomen when a person is standing returns to the thoracic cavity when that person lies down.

Determination of the lung volumes can be useful diagnostically in differentiating between two major types of pulmonary disorders—the restrictive diseases and the obstructive diseases. Restrictive diseases like alveolar fibrosis, which reduce the compliance of the lungs, lead to compressed lung volumes, as seen in Fig. 3-3. The increased elastic recoil of the lungs leads to a lower functional residual capacity, a lower total lung capacity, a lower vital capacity, and lower inspiratory and expiratory reserve volumes and may even decrease the residual volume. The tidal volume may also be decreased, with a corresponding increase in breathing frequency, to minimize the work of breathing.

Obstructive diseases such as emphysema and chronic bronchitis cause increased resistance to airflow. Airways may become completely obstructed because of mucous plugs as well as because of the high intrapleural pressures generated to overcome the elevated airways resistance during a forced expiration. This is especially a problem in emphysema, in which destruction of alveolar septa leads to decreased elastic recoil of the alveoli and less radial traction, which help to hold small airways open. For these reasons the residual volume, the functional residual capacity, and the total lung capacity may be greatly increased in obstructive diseases, as

Figure 3-3 Illustration of typical alterations in the lung volumes and capacities in restrictive and obstructive diseases.

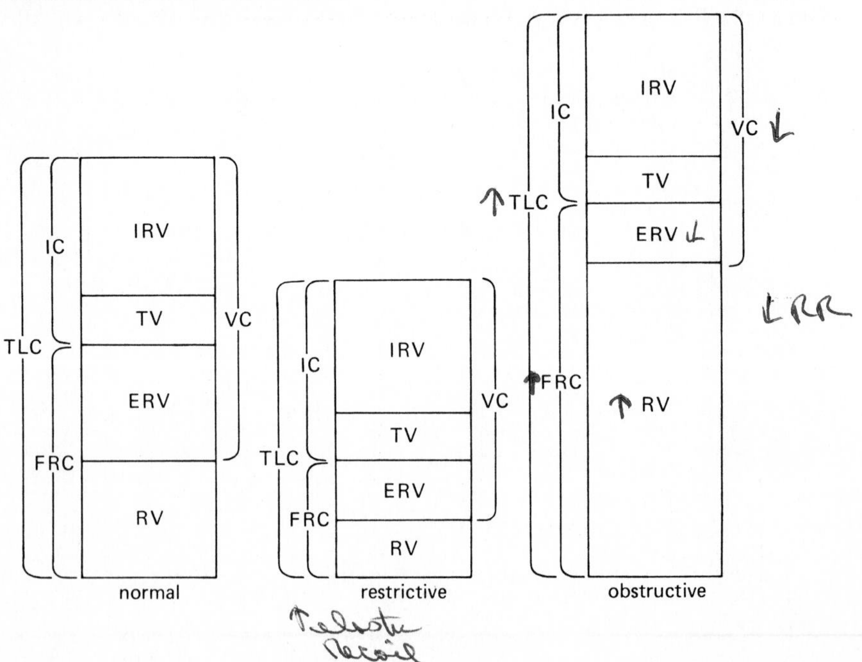

seen in Fig. 3-3. The vital capacity and expiratory reserve volume are usually decreased. The breathing frequency may be decreased to decrease the work expended on overcoming the airways resistance, with a corresponding increase in the tidal volume.

Spirometry

The spirometer is a simple device for measuring gas volumes. The frequently used water spirometer, shown in Fig. 3-4, consists of an inverted

Figure 3-4 Determination of lung volumes and capacities with a spirometer. *A*. Schematic representation of a water-filled spirometer. *B*. Determination of the tidal volume, vital capacity, inspiratory capacity, inspiratory reserve volume, and expiratory reserve volume from a spirometer trace.

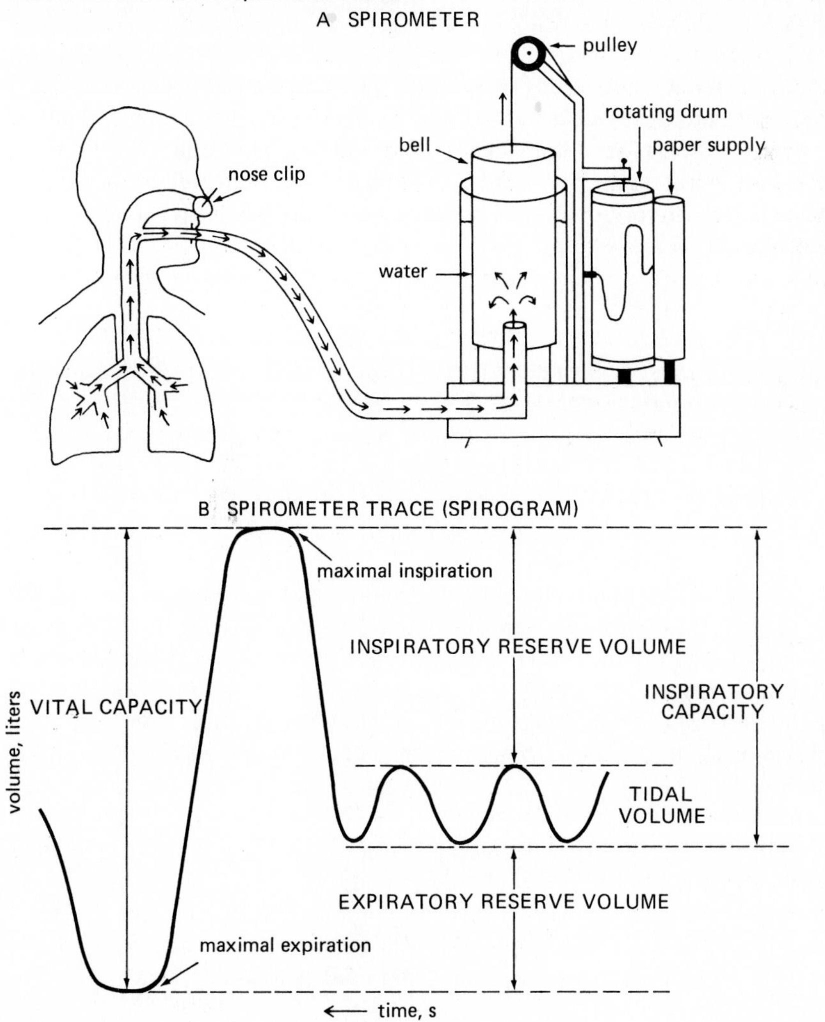

canister or "bell" floating in a water-filled space between two drums. The space inside the inner drum, which is closed off from the atmosphere by the bell, is connected to the subject's lungs by tubing extending to a mouthpiece. As the subject breathes in and out, gas enters and leaves the spirometer and the bell floats higher (during the subject's expiration) and lower (during inspiration). The top of the bell is connected by a pulley to a pen, which then writes on a rotating drum, thus tracing the subject's breathing pattern.

As is evident from Fig. 3-4, the spirometer can measure only the lung volumes that the subject can exchange with it. The subject must be conscious and cooperative. The tidal volume, inspiratory reserve volume, expiratory reserve volume, inspiratory capacity, and the vital capacity can all be measured with a spirometer (as can the FEV_1, FVC, and $FEF_{25-75\%}$, as discussed in Chap. 2). The residual volume, the functional residual capacity, and the total lung capacity, however, cannot be determined with a spirometer because the subject cannot exhale all of the gas in the lungs. The gas in a spirometer is at ambient temperature, pressure, and water vapor saturation and the volumes of gas collected in a spirometer must be converted to equivalent volumes in the body, as shown in Study Quest. 6.

Measurement of Lung Volumes Not Measurable With Spirometry

The lung volumes not measurable with spirometry can be determined by the nitrogen washout technique, by the helium dilution technique, and by body plethysmography. The functional residual capacity is usually determined and the residual volume (which is equal to FRC minus ERV) and the total lung capacity (which is equal to VC plus RV) are then calculated from volumes obtained by spirometry.

Nitrogen-washout Technique In the nitrogen-washout technique the subject breathes 100% oxygen through a one-way valve so that all of the *expired* gas is collected. The concentration of nitrogen in the subject's expired air is monitored with a nitrogen analyzer until it reaches zero. At this point all of the nitrogen is washed out of the subject's lungs. The total volume of *all* the gas the subject expired is determined and this amount is multiplied by the percentage of nitrogen in the mixed expired air, which can be determined with the nitrogen analyzer. The total volume of nitrogen in the subject's lungs *at the beginning of the test* can thus be determined. Since nitrogen constitutes about 80 percent of the subject's initial lung volume, multiplying the initial nitrogen volume by 1.25 gives the subject's initial lung volume. If the test is *begun* at the end of a normal tidal expiration, the volume determined is the functional residual capacity:

Total volume expired $\times$ % N_2 =
original volume of N_2 in lungs
Original volume of N_2 in lungs $\times$ 1.25 = original lung volume

Helium-dilution Technique The helium-dilution technique makes use of the following relationship: If the total *amount* of a substance dissolved in a volume is known and its *concentration* can be measured, the *volume* in which it is dissolved can be determined. For example, if a known amount of a solute is dissolved in an unknown volume of solvent, on condition that the concentration of the solute can be determined, then the volume of solvent can be calculated:

Amount of solute (mg) = concentration of solute (mg/ml)
$\times$ volume of solvent (ml)

In the helium-dilution technique, helium is dissolved in the gas in the lungs and its concentration is determined by means of a helium meter, allowing calculation of the lung volume. Helium is used for this test because it is not taken up by the pulmonary capillary blood, nor does any diffuse out of the blood, and so the total *amount* of helium does not change during the test. The subject breathes in and out of a spirometer filled with a mixture of helium and oxygen, as shown in Fig. 3-5. The helium concentration is monitored continuously with a helium meter until the helium concentration of the subject's inspired air and the concentration of the subject's expired air are identical. At this point the concentration of helium is the same in the subject's lungs as it is in the spirometer, and the test is *stopped* at the end of a normal tidal expiration, i.e., at the functional residual capacity.

The functional residual capacity can then be determined by the following formula:

Total amount of He before test = total amount of He at end of test

$$F_{HE_i} V_{sp_i} = F_{HE_f}(V_{sp_f} + V_{L_f})$$

That is, the *total* amount of helium in the system initially is equal to its initial fractional concentration (F_{HE_i}) times the initial volume of the spirometer (V_{sp_i}). This must be equal to the *total* amount of helium in the lungs and the spirometer at the end of the test, which is equal to the final fractional concentration of helium (F_{HE_f}) times the final volume of the spirometer (V_{sp_f}) *and* the volume of the lungs at the end of the test (V_{L_f}). Since it may take several minutes for the helium concentration to equilibrate between the lungs and the spirometer, in practice, CO_2 is absorbed

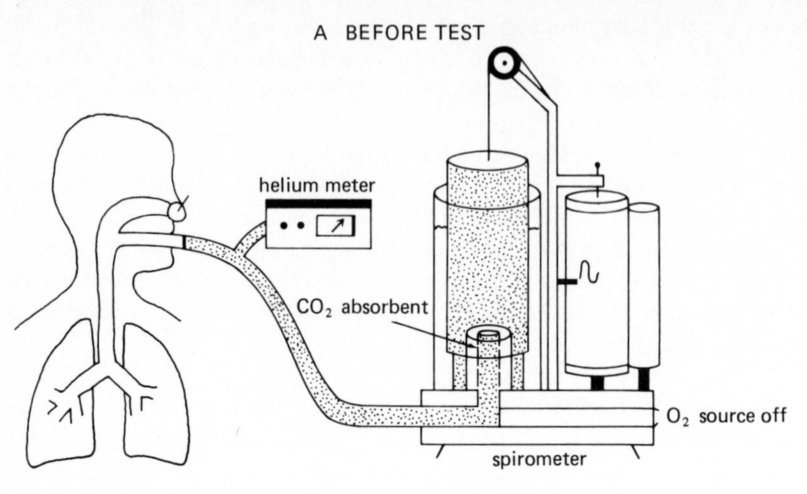

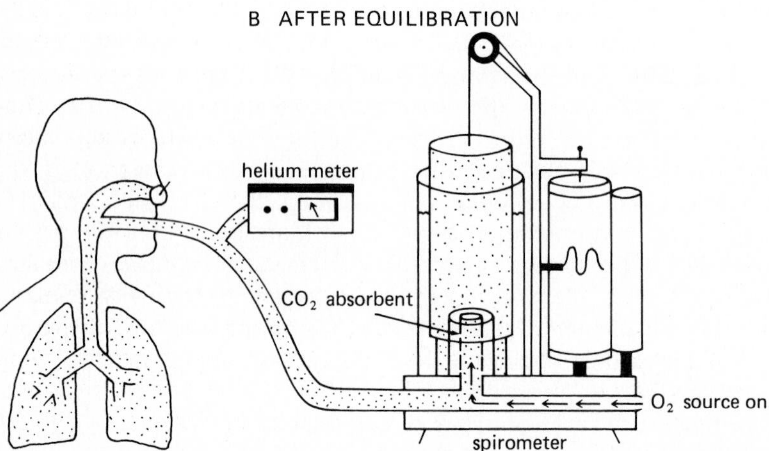

Figure 3-5 The helium dilution technique for the determination of the functional residual capacity. *A.* Before the test, the spirometer is filled with a mixture of helium (denoted by the dots) and oxygen. The concentration of helium is determined by the helium meter. *B.* The subject breathes from the spirometer until the helium concentration in his lungs equilibrates with that in the spirometer. During the equilibration period the subject's expired carbon dioxide is absorbed and oxygen is added to the spirometer at the subject's oxygen consumption rate. The helium concentration and spirometer volume are determined after equilibration, when the subject is at his functional residual capacity.

from the system and oxygen is added to the spirometer at the rate it is used by the subject. Both the nitrogen-washout and helium-dilution methods can be used on unconscious patients.

Body Plethysmography A problem common to both the nitrogen-washout technique and the helium-dilution technique is that neither can measure trapped gas. Furthermore, if the patient's lungs have many alveoli served by airways with high resistance to airflow (the ''slow alveoli'' discussed at the end of Chap. 2), it may take a very long time for all of the nitrogen to wash out of the patient's lungs or for the inspired and expired helium concentrations to equilibrate. In such patients, measurements of the lung volumes with a body plethysmograph are much more accurate because they do include trapped gas.

The body plethysmograph makes use of Boyle's law, which states that for a closed container at a constant temperature, the pressure times the volume is constant. The body plethysmograph, an expensive piece of equipment, is shown schematically in Fig. 3-6.

As can be seen from the figure, the body plethysmograph is an airtight chamber large enough to hold the subject. The subject sits in the closed plethysmograph or ''box'' and breathes through a mouthpiece and tubing. The tubing contains a sidearm connected to a pressure transducer (''mouth pressure''); an electrically controlled shutter that can occlude the airway when activated by the person conducting the test; and a pneumotachograph to measure airflow, allowing the operator to follow the subject's breathing pattern. A second pressure transducer, which must be very sensitive, monitors the pressure in the plethysmograph (''box pressure'').

After the subject breathes through the open tube for a while, to establish a normal breathing pattern, the operator closes the shutter in the airway at the end of a normal tidal expiration. At this point the subject *breathes in* for an instant against a closed airway. As the subject breathes in against the closed airway, the chest continues to expand and the pressure measured by the transducer in the plethysmograph (P_{box}) *increases* because the *volume* of air in the plethysmograph (V_{box}) decreases by the amount the patient's chest volume increased (ΔV):

$$P_{box_i} \times V_{box_i} = P_{box_f} \times (V_{box_i} - \Delta V) \qquad (1)$$

where $(V_{box_i} - \Delta V) = V_{box_f}$

That is, the product of the initial box pressure times the initial box volume must equal the final box pressure times the final box volume (the initial box volume minus a change in volume), according to Boyle's law. Of course, direct measurement of box volume, which is really equal to the volume of the plethysmograph *minus the volume occupied by the patient,* is impossible, and so the plethysmograph is calibrated with the patient in it by injecting known volumes of air into the plethysmograph and determining the increase in pressure. After such a graph of pressure changes with known changes in volume has been constructed, the ΔV in Eq. (1) can be determined.

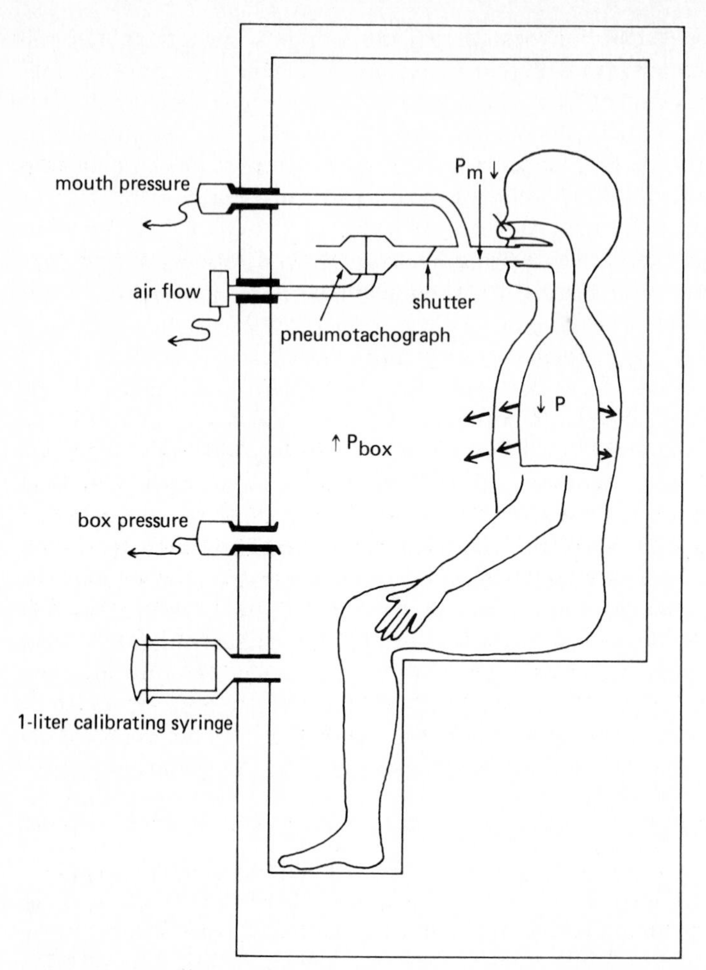

Figure 3-6 The use of the body plethysmograph for the determination of the functional residual capacity. The subject is seated in the small airtight chamber and breathes through the apparatus shown. By monitoring the subject's airflow with a pneumotachograph, the operator can briefly occlude the subject's airway at end expiration. As the subject makes an inspiratory effort against the closed airway, the pressure in the chamber (P_{box}) increases and the pressure at the subject's mouth (P_M) decreases. The subject's functional residual capacity can then be calculated.

The product of the pressure measured at the mouth (P_M) times the volume of the patient's lungs (V_L) must also be constant during the inspiration against a closed airway. As the patient breathes in, the volume of the lungs increases by the same amount as the decrease in the volume of the box determined in Eq. (1) above (ΔV). As the lung volume increases, the pressure measured at the mouth decreases, as predicted by Boyle's law:

$$P_{M_i} \times V_{L_i} = P_{M_f} \times (V_{L_i} + \Delta V) \tag{2}$$

where $(V_{L_i} + \Delta V) = V_{L_f}$

The ΔV in Eq. (2) is equal to that solved for in Eq. (1) and V_{L_i} is now solved for. It is the functional residual capacity, since the airway was occluded at the end of a normal tidal expiration. In current practice all of the calculations described above are made automatically by a computer receiving inputs from the pressure transducers.

ANATOMIC DEAD SPACE AND ALVEOLAR VENTILATION

The volume of air entering and leaving the nose or mouth per minute, the *minute volume,* is not equal to the volume of air entering and leaving the alveoli per minute. *Alveolar ventilation* is *less* than the minute volume because the last part of each inspiration remains in the conducting airways and does not reach the alveoli. Similarly, the last part of each expiration remains in the conducting airways and is not expelled from the body. No gas exchange occurs in the conducting airways for anatomic reasons: The walls of the conducting airways are too thick for much diffusion to take place; mixed venous blood does not come into contact with the air. The conducting airways are therefore referred to as the *anatomic dead space.*

The anatomic dead space is illustrated in Fig. 3-7. A subject breathes in from a balloon filled with 500 ml of a test gas that is not taken up by or

Figure 3-7 Illustration of the anatomic dead space. *A.* The subject inspires 500 ml from a balloon filled with a high concentration of a test gas (denoted by the dots). *B.* At the end of the inspiration, only 350 ml of the test gas has reached the alveoli. This 350 ml is added to the 2 to 3 liters of alveolar gas already in the lungs at the FRC, and so its concentration is diluted. The other 150 ml of test gas remains virtually unchanged in the subject's anatomic dead space. *C.* At end expiration diluted test gas remains equally concentrated in alveolar air and the anatomic dead space. The test gas in the balloon is a mixture of undiluted gas from the dead space and diluted alveolar gas.

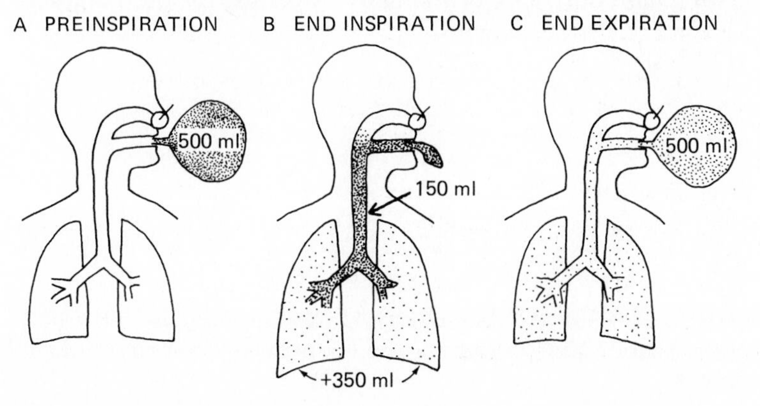

liberated from the pulmonary capillary blood, for example, helium. Initially (Fig. 3-7*A*) there is no test gas in the subject's airways or lungs. The subject then (Fig. 3-7*B*) breathes in all 500 ml of the gas. Not all of the gas, however, reaches the alveoli. The final portion of the inspired gas remains in the conducting airways, completely filling them. The volume of the test gas reaching the alveoli is equal to the volume breathed in from the balloon minus the volume of the anatomic dead space, in this case 500 ml − 150 ml, or 350 ml. The 350 ml of test gas mixes with the air already in the alveoli and it is diluted. During expiration (Fig. 3-7*C*) the first gas breathed back into the balloon is the undiluted test gas that remained in the anatomic dead space. Following the undiluted test gas is part of the gas that reached the alveoli and was diluted by the alveolar air. The last 150 ml of alveolar gas breathed out remains in the anatomic dead space. The concentration of test gas collected in the balloon after expiration is lower than it was before the breath but is higher than the concentration left in the alveoli and conducting airways because it is composed of pure test gas from the anatomic dead space and diluted test gas from the alveoli.

Thus, for any respiratory cycle, not all of the tidal volume reaches the alveoli because the last part of each inspiration and each expiration remains in the dead space. The relationship between the tidal volume (V_T) breathed in and out through the nose or mouth, the dead space volume (V_D), and the volume of gas entering and leaving the alveoli per breath (V_A) is

$$V_T = V_D + V_A$$

or

$$V_A = V_T - V_D$$

Thus, if a person with an anatomic dead space of 150 ml has a tidal volume of 500 ml per breath, then only 350 ml of gas enters and leaves the alveoli per breath.

By multiplying both sides of the above equation by the breathing frequency (n) in breaths per minute,

$$n(V_A) = n(V_T) - n(V_D)$$

Thus, if $n = 12$ breaths per minute in the example above,

$$4200 \frac{\text{ml}}{\text{min}} = 6000 \frac{\text{ml}}{\text{min}} - 1800 \frac{\text{ml}}{\text{min}}$$

The alveolar ventilation ($\dot{V}_A$) in liters per minute is equal to the minute volume ($\dot{V}_E$) minus the volume wasted ventilating the dead space per minute ($\dot{V}_D$):

$$\dot{V}_A = \dot{V}_E - \dot{V}_D$$

The dots over the V's indicate *per minute*. The symbol $\dot{V}_E$ is used because expired gas is usually collected. There is a difference between the volume of gas inspired and the volume of gas expired because as air is inspired it is heated to body temperature and humidified and also because normally less carbon dioxide is produced than oxygen is consumed.

MEASUREMENT OF ALVEOLAR VENTILATION

Alveolar ventilation cannot be measured directly but must be determined from the tidal volume, the breathing frequency, and the dead space ventilation, as noted in the previous section.

Measurement of Anatomic Dead Space

For a normal, healthy subject, the anatomic dead space can be estimated by referring to a table of standard values matched to sex, age, height, and weight or body surface area. A reasonable estimate of anatomic dead space is 1 ml of dead space per pound of body weight. Nevertheless, it may be important to determine the anatomic dead space in a particular patient. This can be done by using Fowler's method. This method uses a nitrogen meter to analyze the expired nitrogen concentration after a single inspiration of 100% oxygen. The expired gas volume is measured simultaneously. Fowler's method is summarized in Fig. 3-8.

The subject breathes in a single breath of 100% oxygen through a one-way valve, holds it in for a second, and then exhales it through the one-way valve. *Nitrogen concentration* at the mouth and the *volume expired* are monitored simultaneously. Initially the nitrogen concentration at the mouth is 80%, that of the ambient atmosphere. As the stopcock is turned and the subject begins to inspire 100% oxygen, the nitrogen concentration at the mouth falls to zero. The subject holds his or her breath for a second or so and then exhales through the valve into a spirometer or pneumotachograph. The first part of the expired air registers 0% nitrogen because it is undiluted 100% oxygen from the anatomic dead space. In the transitional period that follows, the expired gas registers a slowly rising nitrogen concentration. During this time the expired air is a mixture of dead space gas and alveolar gas because of a gradual transition between the conducting pathways and the respiratory bronchioles, as was seen in Fig. 1-5. The final portion of expired gas comes solely from the alveoli and is called the *alveolar plateau*. Its nitrogen concentration is less than 80 percent because some of the breath of 100% oxygen reached the alveoli and diluted the alveolar nitrogen concentration, as shown in Fig. 3-7. The volume of the anatomic dead space is the volume expired between the

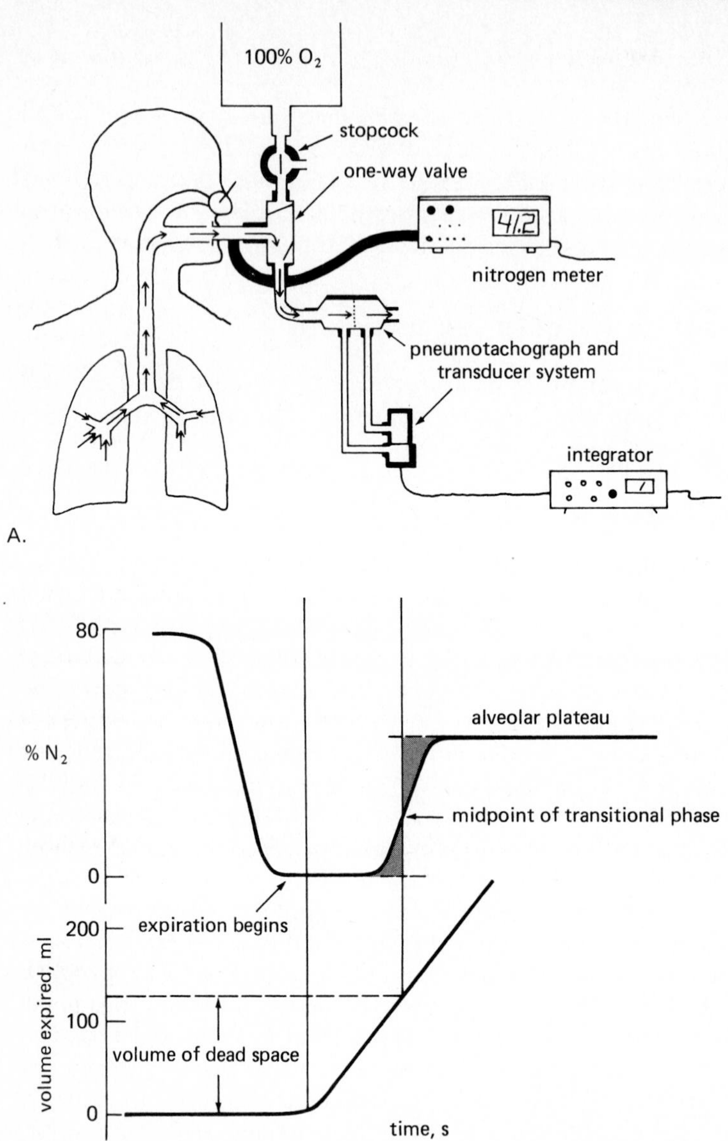

Figure 3-8 Fowler's method for the determination of anatomic dead space. *A*. The subject takes a single breath of 100% oxygen, holds his breath for a second, and then exhales. Nitrogen concentration is monitored along with the volume of gas expired, in this case by integrating with time the airflow (liters/s) determined by a pneumotachograph–differential air pressure transducer system. *B*. The volume of gas expired between the beginning of the exhalation and the midpoint of the rising phase of the expired nitrogen concentration trace is the anatomic dead space. (The midpoint is determined such that the two shaded areas are equal.)

beginning of the expiration and the midpoint of the transitional phase, as shown on Fig. 3-8.

Physiological Dead Space: The Bohr Equation

Fowler's method is especially useful for the determination of the *anatomic* dead space. It does not, however, permit the calculation of another form of wasted ventilation in the lung—the *alveolar dead space*. The alveolar dead space is the volume of gas that enters *unperfused* alveoli per breath. No gas exchange occurs in these alveoli for physiological reasons, not for reasons of anatomy. A healthy person has little or no alveolar dead space, but a person with a low cardiac output might have a great deal of alveolar dead space, for reasons explained in the next chapter.

The Bohr equation permits the computation of both the anatomic and the alveolar dead space. The anatomic dead space *plus* the alveolar dead space is known as the *physiological dead space:*

Physiologic dead space = anatomic dead space + alveolar dead space

The Bohr equation makes use of a simple concept: Any measurable volume of carbon dioxide found in the mixed *expired* gas must come from alveoli that are *both ventilated and perfused* because there are negligible amounts of carbon dioxide in inspired air. Inspired air remaining in the anatomic dead space or entering unperfused alveoli will leave the body as it entered (except for having been heated to body temperature and humidified), contributing little or no carbon dioxide to the mixed expired air:

$$\underbrace{F_{E_{CO_2}} \times V_T}_{\text{The volume of } CO_2 \text{ in mixed expired air}} = \underbrace{F_{I_{CO_2}} \times V_{D_{CO_2}}}_{\text{The volume of } CO_2 \text{ coming from dead space}} + \underbrace{F_{A_{CO_2}} \times V_A}_{\text{The volume of } CO_2 \text{ coming from alveoli}}$$

where F = fractional concentration
E = expired
I = inspired
A = alveolar
$V_{D_{CO_2}}$ = dead space for CO_2 (physiological dead space)
$F_{A_{CO_2}}$ = fractional concentration of CO_2 in alveoli that are both ventilated and perfused

Since $F_{I_{CO_2}}$ is approximately equal to zero, the $F_{I_{CO_2}} \times V_{D_{CO_2}}$ term drops out. Substituting $(V_T - V_{D_{CO_2}})$ for V_A,

$$F_{E_{CO_2}} \times V_T = F_{A_{CO_2}}(V_T - V_{D_{CO_2}})$$
$$F_{E_{CO_2}} \times V_T = F_{A_{CO_2}} \times V_T - F_{A_{CO_2}} \times V_{D_{CO_2}}$$
$$V_{D_{CO_2}} \times F_{A_{CO_2}} = V_T(F_{A_{CO_2}} - F_{E_{CO_2}})$$
$$\frac{V_{D_{CO_2}}}{V_T} = \frac{F_{A_{CO_2}} - F_{E_{CO_2}}}{F_{A_{CO_2}}}$$

Since

$$F_{CO_2} = \frac{P_{CO_2}}{P_{Total}}$$

then $$\frac{V_{D_{CO_2}}}{V_T} = \frac{P_{A_{CO_2}} - P_{E_{CO_2}}}{P_{A_{CO_2}}}$$

The P_{CO_2} of the collected mixed expired gas can be determined with a CO_2 meter. The CO_2 meter is often used to *estimate* the alveolar P_{CO_2} by analyzing the gas exhaled at the end of a normal tidal expiration ("end-tidal CO_2"). But in a person with significant alveolar dead space, the estimated alveolar P_{CO_2} obtained in this fashion may be quite inaccurate because some of this end-tidal gas comes from unperfused alveoli. There is, however, an equilibrium between the P_{CO_2} of *perfused* alveoli and end-capillary P_{CO_2} (as will be discussed in Chap. 6), so that in patients without significant venous-to-arterial shunts, the *arterial* P_{CO_2} represents the mean P_{CO_2} of the *perfused* alveoli. Therefore the Bohr equation can be rewritten as

$$\frac{V_{D_{CO_2}}}{V_T} = \frac{Pa_{CO_2} - P_{E_{CO_2}}}{Pa_{CO_2}}$$

The tidal volume is determined with a spirometer and the physiological dead space is then calculated. If the arterial P_{CO_2} is greater than the alveolar P_{CO_2} determined by sampling the end-tidal CO_2, then the physiological dead space is likely greater than the anatomic dead space; i.e., there is significant alveolar dead space. Situations in which alveoli are ventilated but not perfused include those in which portions of the pulmonary vasculature have been occluded by blood clots in the venous blood (pulmonary emboli), situations in which there is low venous return leading to low right ventricular output (hemorrhage), or situations in which alveolar pressure is extremely high (positive-pressure ventilation with positive end-expiratory pressure).

The anatomic dead space can be altered by bronchoconstriction, which decreases V_D; bronchodilation, which increases V_D; or by traction or compression of the airways, which increase and decrease V_D, respectively.

ALVEOLAR VENTILATION AND ALVEOLAR OXYGEN AND CARBON DIOXIDE LEVELS

The levels of oxygen and carbon dioxide in alveolar gas are determined by the alveolar ventilation, the oxygen consumption ($\dot{V}O_2$) of the body, and the carbon dioxide production by the body ($\dot{V}CO_2$). Each breath brings into the 3 liters of gas already in the lungs approximately 350 ml of fresh air containing about 21% oxygen and removes about 350 ml of air containing about 5 to 6% carbon dioxide. Meanwhile, about 250 ml of carbon dioxide per minute diffuses from the pulmonary capillary blood into the alveoli and about 300 ml of oxygen per minute diffuses from the alveolar air into the pulmonary capillary blood.

Partial Pressures of Respiratory Gases

According to Dalton's law, in a gas mixture, the pressure exerted by each individual gas in a space is independent of the pressures of other gases in the mixture. The partial pressure of a particular gas is equal to its fractional concentration times the total pressure of all the gases in the mixture. Thus for any gas in a mixture (gas_1) its partial pressure

$$P_{gas_1} = \% \text{ of total gas} \times P_{total}$$

Oxygen constitutes 20.93% of dry atmospheric air. At a standard barometric pressure of 760 mmHg

$$PO_2 = 0.2093 \times 760 \text{ torr} = 159 \text{ torr}$$

(The units mmHg are expressed as torr, in honor of Evangelista Torricelli, the inventor of the barometer.) Carbon dioxide constitutes only about 0.04% of dry atmospheric air, and so

$$PCO_2 = 0.0004 \times 760 \text{ torr} = 0.3 \text{ torr}$$

Dry Atmospheric Gas at Standard Barometric Pressure	
PO_2	159.0 torr
PCO_2	0.3 torr
PN_2	600.6 torr

As air is inspired through the upper airways, it is heated and humidified, as will be discussed in Chap. 10. The partial pressure of water vapor is a relatively constant 47 torr at body temperature, and so the humidification of a liter of dry gas *in a container* at 760 torr would increase its total pressure to 760 torr + 47 torr = 807 torr. In the body the gas will simply expand, according to Boyle's law, so that 1 liter of gas at 760 torr is *diluted* by the added water vapor. The P_{O_2} of inspired air (saturated with water vapor at a standard barometric pressure) then is equal to

$$0.2093\ (760 - 47)\ \text{torr} = 149\ \text{torr}$$

The P_{CO_2} of inspired air is 0.0004 (760 − 47) torr = 0.29 torr, still too low to be of any interest.

Inspired Gas at Standard Barometric Pressure	
$P_{I_{O_2}}$	149.0 torr
$P_{I_{CO_2}}$	0.3 torr
$P_{I_{N_2}}$	564.0 torr
$P_{I_{H_2O}}$	47.0 torr

Alveolar gas is composed of the 2.5 to 3 liters of gas already in the lungs at the functional residual capacity and the approximately 350 ml per breath entering and leaving the alveoli. About 300 ml of oxygen is continuously diffusing from the alveoli into the pulmonary capillary blood per minute at rest and is being replaced by alveolar ventilation. Similarly, about 250 ml of CO_2 is diffusing from the mixed venous blood in the pulmonary capillaries into the alveoli per minute and is then removed by alveolar ventilation. (The P_{O_2} and P_{CO_2} of mixed venous blood are about 40 torr and 45 to 46 torr, respectively.) Because of these processes, the partial pressures of oxygen and carbon dioxide in the alveolar air are determined by the alveolar ventilation, the oxygen consumption, and the carbon dioxide production. As we will see in Chap. 9, alveolar ventilation is normally adjusted by the respiratory control center in the brain to keep mean arterial and alveolar P_{CO_2} at about 40 torr. Mean alveolar P_{O_2} is about 104 torr.

Alveolar Gas at Standard Barometric Pressure	
$P_{A_{O_2}}$	104 torr
$P_{A_{CO_2}}$	40 torr
$P_{A_{N_2}}$	569 torr
$P_{A_{H_2O}}$	47 torr

The alveolar P_{O_2} increases by 2 to 4 torr with each normal tidal inspiration and falls slowly until the next inspiration. Similarly the alveolar P_{CO_2} falls 2 to 4 torr with each inspiration and increases slowly until the next inspiration. Expired air is a mixture of about 350 ml of alveolar air and 150 ml of air from the dead space. Therefore the P_{O_2} of mixed expired air is higher than alveolar P_{O_2} and less than the inspired P_{O_2}, or approximately 120 torr. Similarly, the P_{CO_2} of mixed expired air is much higher than the inspired P_{CO_2} but lower than the alveolar P_{CO_2}, or about 27 torr.

Mixed Expired Air at Standard Barometric Pressure	
$P_{E_{O_2}}$	120 torr
$P_{E_{CO_2}}$	27 torr
$P_{E_{N_2}}$	566 torr
$P_{E_{H_2O}}$	47 torr

Alveolar Ventilation and Carbon Dioxide

The concentration of carbon dioxide in the alveolar gas is, as already discussed, dependent on the alveolar ventilation and on the rate of carbon dioxide production by the body (and its delivery to the lung in the mixed venous blood). The volume of carbon dioxide expired per unit of time ($\dot{V}E_{CO_2}$) is equal to the alveolar ventilation ($\dot{V}A$) times the alveolar fractional concentration of CO_2 (FA_{CO_2}). No carbon dioxide comes from the dead space.

$$\dot{V}E_{CO_2} = \dot{V}A \times FA_{CO_2}$$

Similarly, the fractional concentration of carbon dioxide in the alveoli is directly proportional to the carbon dioxide production by the body ($\dot{V}_{CO_2}$) and inversely proportional to the alveolar ventilation:

$$FA_{CO_2} \propto \frac{\dot{V}_{CO_2}}{\dot{V}A}$$

Since $FA_{CO_2} \times (760 - 47) = PA_{CO_2}$

then $PA_{CO_2} \propto \dfrac{\dot{V}_{CO_2}}{\dot{V}A}$

In healthy people alveolar P_{CO_2} is in equilibrium with arterial P_{CO_2} (Pa_{CO_2}). Thus, if alveolar ventilation is doubled (and carbon dioxide production is unchanged), then the alveolar and arterial P_{CO_2} are reduced by one-half. If alveolar ventilation is cut in half, near 40 torr, then alveolar and arterial P_{CO_2} will double. This can be seen in the upper part of Fig. 3-9. The curve is not linear because CO_2 *production* increases at high levels of alveolar ventilation.

Alveolar Ventilation and Oxygen

It is evident that as alveolar ventilation increases, the alveolar P_{O_2} will also increase. Doubling alveolar ventilation, however, cannot double PA_{O_2}

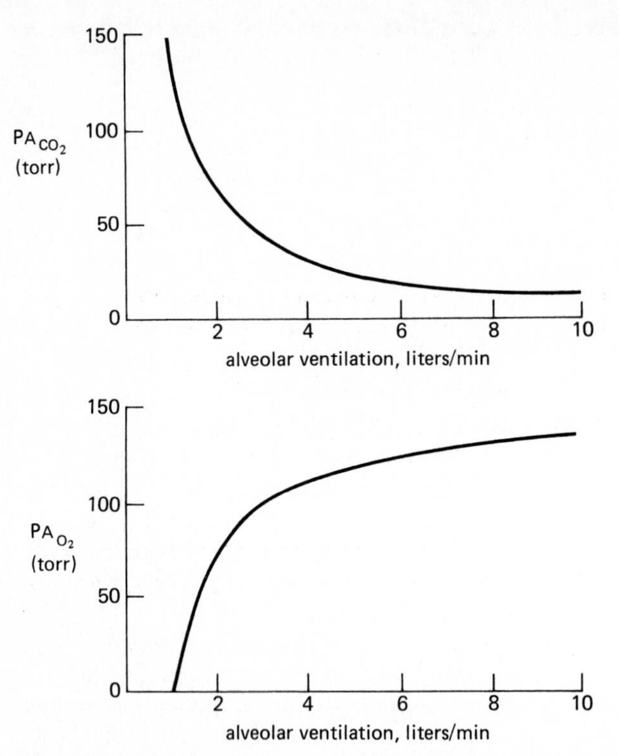

Figure 3-9 Predicted alveolar gas tensions for different levels of alveolar ventilation. *(After Nunn, 1977. Reproduced with permission.)*

in a person whose alveolar Po_2 is already 104 torr because the highest PA_{O_2} one can achieve (breathing air at sea level) is the inspired Po_2 of about 149 torr. The alveolar Po_2 can be calculated by using the alveolar air equation. (The derivation of this formula is outside of the scope of this book.)

Alveolar Air Equation

$$PA_{O_2} = PI_{O_2} - \frac{PA_{CO_2}}{R} + F$$

where R = respiratory exchange ratio $\frac{\dot{V}CO_2}{\dot{V}O_2}$

F = a correction factor

As alveolar ventilation increases, the alveolar Pco_2 decreases, bringing the alveolar Po_2 closer to the inspired Po_2, as can be seen in the lower part of Fig. 3-9.

REGIONAL DISTRIBUTION OF ALVEOLAR VENTILATION

In this chapter we have seen that a 70-kg person has about 2.5 to 3 liters of gas in his or her lungs at the functional residual capacity. Each breath brings about 350 ml of fresh gas into the alveoli and removes about 350 ml of alveolar air from the lung. Although it is reasonable to assume that the alveolar ventilation is distributed fairly evenly to alveoli throughout the lungs, this is not the case. Studies performed on normal subjects seated upright have shown that alveoli in the lower regions of the lungs receive more ventilation per unit volume than those in the upper regions of the lung.

Demonstration of Differences Between Dependent and Nondependent Regions

If a normal subject, seated in the upright posture and breathing normally (inspiring from the FRC), takes a single breath of a mixture of air and radioactive xenon 133, the relative ventilation of various regions of the lung can be determined by placing scintillation counters over appropriate areas of the thorax, as shown at left in Fig. 3-10.

It is assumed that if the air and xenon 133 are well mixed, then the amount of radioactivity measured by the scintillation counters in each region will be directly proportional to the relative ventilation (the ventilation per unit volume) in each region.

The results of a series of such experiments are shown on the graph on the right side of Fig. 3-10. In a subject seated in the upright posture and breathing normally from the functional residual capacity, the lower re-

Figure 3-10 Regional distribution of alveolar ventilation as determined by a breath of a mixture of ^{133}Xe and O_2. *(Data of Bryan, 1964. Reproduced with permission.)*

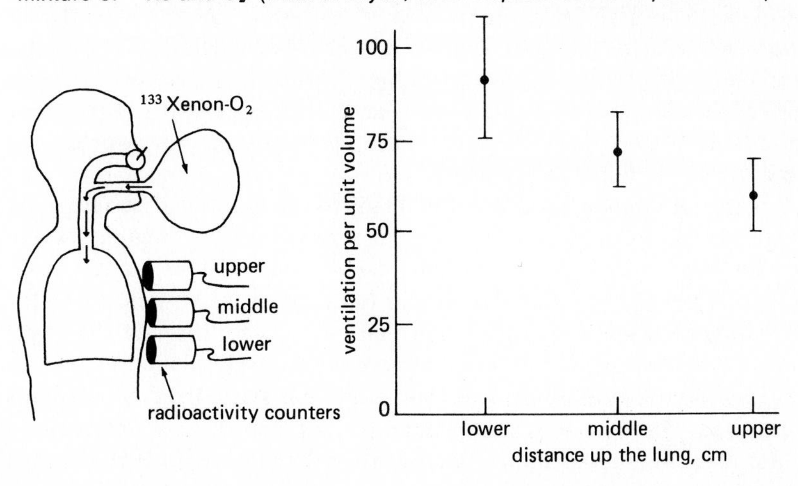

gions of the lung are relatively better ventilated than the upper regions of the lung.

If a similar study is done on a *supine* subject, the regional differences in ventilation between the *anatomic* upper, middle, and lower regions of the lung disappear, although there is better relative ventilation of the dorsal regions of the lungs than that of the ventral regions of the lungs. The regional differences in ventilation thus appear to be influenced by gravity, with regions of the lung lower with respect to gravity (the *"dependent"* regions) relatively better ventilated than those regions above them (the *"nondependent"* regions).

Explanation for Differences in Regional Alveolar Ventilation

In Chap. 2, the intrapleural surface pressure was discussed as if it were uniform throughout the thorax. Precise measurements made of the intrapleural surface pressures of intact chests in the upright position have shown that this is not the case: The intrapleural surface pressure is *less negative* in the lower, gravity-dependent regions of the thorax than it is in the upper, nondependent regions. There is a gradient of the intrapleural surface pressure such that for every centimeter of vertical displacement down the lung (from nondependent to dependent regions) the intrapleural surface pressure increases by about +0.2 cmH_2O. This gradient is apparently caused by gravity and by mechanical interactions between the lung and the chest wall.

The influence of this gradient of intrapleural surface pressure on regional alveolar ventilation can be explained by predicting its effect on the transpulmonary pressure in upper and lower regions of the lung. As shown at left in Fig. 3-11, at the FRC *alveolar* pressure is zero in both regions of the lung. Since the intrapleural pressure is *more negative in upper regions* of the lung than it is in lower regions of the lung, the *transpulmonary pressure* (alveolar minus intrapleural) is greater in upper regions of the lung than it is in lower regions of the lung. Because the alveoli in upper regions of the lung are subjected to greater distending pressures than those in more dependent regions of the lung, they have greater *volumes* than the alveoli in more dependent regions.

It is this difference in *volume* that leads to the difference in *ventilation* between alveoli located in dependent and nondependent regions of the lung. This can be seen on the hypothetical pressure-volume curve shown at right in Fig. 3-11. This curve is similar to the pressure-volume curve for a whole lung shown in Fig. 2-6, except that this curve is drawn for a single alveolus. The abscissa is the transpulmonary pressure (alveolar pressure minus intrapleural pressure). The ordinate is the volume of the alveolus expressed as a percent of its maximum.

The alveolus in the upper, nondependent region of the lung has a

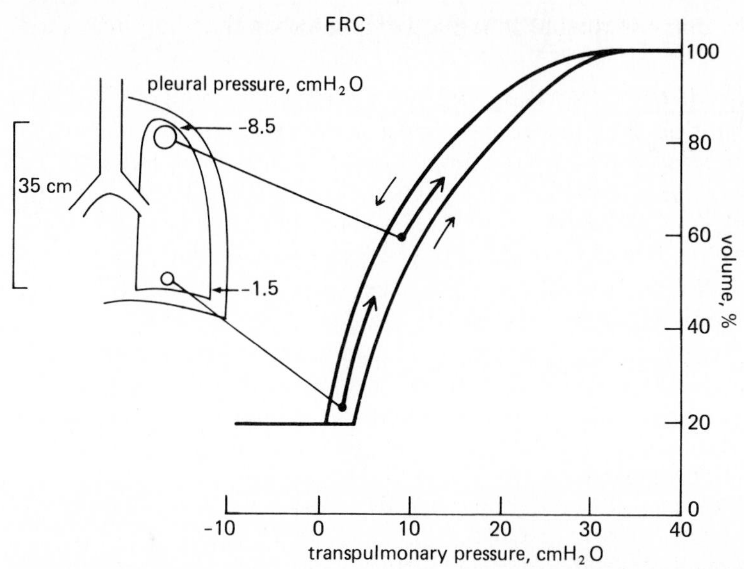

Figure 3-11 Effect of the pleural surface pressure gradient on the distribution of inspired gas at the *functional residual capacity.* (*After Milic-Emili, 1977.*)

larger transpulmonary pressure than the alveolus in a more dependent region because the intrapleural pressure in the upper, nondependent regions of the lung is more negative than it is in more dependent regions. Because of this greater transpulmonary pressure, the alveolus in the upper region of the lung has a greater volume than the alveolus in a more gravity-dependent region of the lung. At the functional residual capacity, the alveolus in the upper part of the lung is on a *less steep portion* of the alveolar pressure-volume curve (that is, it is *less compliant*) in Fig. 3-11 than is the alveolus in the lower region of the lung. Therefore, any change in the transpulmonary pressure during a normal respiratory cycle will cause a greater *change in volume* in the alveolus in the lower, gravity-dependent region of the lung than it will in the alveolus in the nondependent region of the lungs, as shown by the arrows in the figure. Because the alveoli in the lower parts of the lung have a greater change in volume per inspiration and per expiration, they are better *ventilated* than those alveoli in nondependent regions (during eupneic breathing from the FRC).

A second effect of the intrapleural pressure gradient in a person seated upright is on regional static lung volume, as is evident from the above discussion. At the functional residual capacity, most of the alveolar air is in upper regions of the lung because those alveoli have larger volumes. Most of the expiratory reserve volume is also in upper portions of the lung. On the other hand most of the inspiratory reserve volume and inspiratory capacity are in lower regions of the lung.

Alterations of Distribution at Different Lung Volumes or With Age

As discussed in the previous section, most of the air inspired during a tidal breath begun at the functional residual capacity enters the dependent alveoli. If a slow inspiration is begun at the *residual volume,* however, the initial part of the breath (inspiratory volume less than the ERV) enters the nondependent upper alveoli, and the dependent alveoli begin to fill later in the breath. The intrapleural pressure gradient from the upper parts of the lung to the lower parts of the lung is also the cause of this preferential ventilation of nondependent alveoli at low lung volumes.

Positive intrapleural pressures are generated by the expiratory muscles during a forced expiration to the residual volume. This results in dynamic compression of small airways, as described in Chap. 2. At the highest intrapleural pressures these airways close and gas is trapped in their alveoli. Because of the gradient of intrapleural pressure found in the upright lung, at low lung volumes the pleural surface pressure is more positive in lower regions of the lung than it is in upper regions. This means that airway closure will occur *first* in airways in lower regions of the lung, as can be seen in the hypothetical alveolar pressure-volume curve at the residual volume shown in Fig. 3-12. The expiratory effort has ended and the inspiratory effort has just begun. Airways in the lowest regions of the lung are still closed and the local pleural surface pressure is still slightly positive. No air enters these alveoli during the first part of the inspiratory

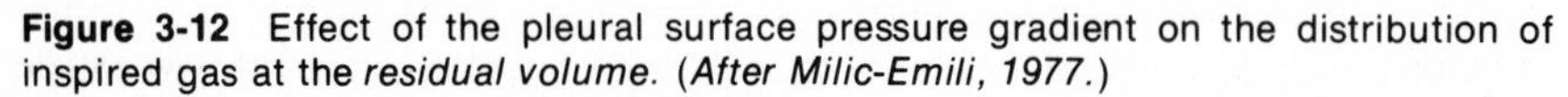

Figure 3-12 Effect of the pleural surface pressure gradient on the distribution of inspired gas at the *residual volume.* (*After Milic-Emili, 1977.*)

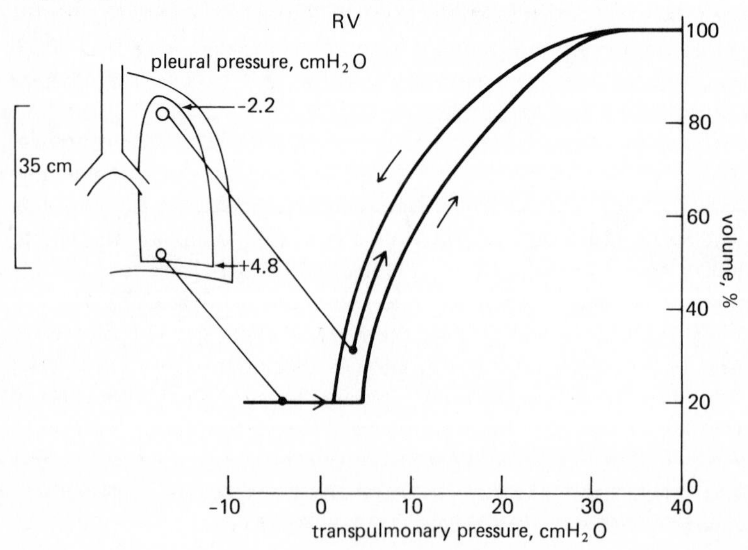

effort (as indicated by the horizontal arrow) until sufficient negative pressure is generated to open these closed airways.

In contrast to the situation at the functional residual capacity, at the residual volume the alveoli in the upper regions of the lungs are now on a much steeper portion of the pressure-volume curve. They now have a much greater change in volume per change in transpulmonary pressure—they are more compliant at this lower lung volume. Therefore they receive more of the air initially inspired from the residual volume.

It is important to note that even at low lung volumes the upper alveoli still are larger in volume than the lower gravity-dependent alveoli. They therefore constitute most of the residual volume.

In older people there is a progressive loss of lung elastic recoil (and chest wall elastic recoil also increases), leading to an elevated functional residual capacity. The loss of alveolar elastic recoil also leads to airway closure at higher lung volumes, as well as to increased residual volumes. Airway closure may occur in dependent airways even at the FRC. Such older persons may therefore have relatively more ventilation of upper airways than a younger person.

Patients with emphysema have greatly decreased alveolar elastic recoil, leading to high functional residual capacities, extremely high residual volumes, and airway closure in dependent parts of the lung even at high lung volumes. They also have relatively more ventilation of nondependent alveoli.

THE CLOSING VOLUME

The lung volume at which airway closure *begins* to occur is known as the *closing volume*. It can be demonstrated by utilizing the same equipment used in Fowler's method for the quantification of the anatomic dead space, seen in Fig. 3-8. This method can also demonstrate certain maldistributions of alveolar ventilation.

The subject, seated upright, *starts from the residual volume* and inspires a single breath of 100% oxygen all the way up to the total lung capacity. He or she then exhales all the way back down to the residual volume. Nitrogen concentration at the mouth and the volume of gas expired are monitored simultaneously throughout the expiration.

Consider what occurs during the *first* expiration to the residual volume. Because of the gradient of intrapleural pressure from the top of the lung to the bottom of the lung, the alveoli in upper parts of the lung are larger than those in lower regions of the lung. Any gas left in the lungs at the end of this initial forced expiration to the residual volume is about 80% nitrogen, and so most of the nitrogen (and most of the RV) are in upper parts of the lung. Alveoli in lower portions of the lung have smaller volumes and thus contain less nitrogen. At the bottom part of the lung airways are closed, trapping whatever small volume of gas remains in these alveoli.

The subject then inspires 100% oxygen to the total lung capacity. Although the initial part of this breath will likely enter the upper alveoli, as described previously, most of this 100% oxygen will enter the more dependent alveoli. (The *very* first part, which does enter the upper alveoli, is dead space gas, which is 80% nitrogen anyway.) If we could measure the nitrogen concentration of alveoli in different parts of the lung at this point, we would find the highest nitrogen concentration in upper regions of the lung and the lowest nitrogen concentration in the lower regions of the lung.

The subject then exhales to the residual volume as the expired nitrogen concentration and gas volume are monitored. The expired nitrogen concentration trace is shown in Fig. 3-13.

The first gas the subject exhales (phase I) is pure gas from the anatomic dead space. It is still virtually 100% oxygen or 0% nitrogen. The second portion of gas exhaled by the subject (phase II) is a mixture of dead-space gas and alveolar gas. The third portion of gas expired by the subject is mixed alveolar gas from the upper and lower regions (phase III or the "alveolar plateau").

Figure 3-13 Expired nitrogen concentration after inhalation of a single breath of 100% O_2 from the residual volume to the total lung capacity. Subject exhales to the residual volume. *Phase I:* 0% nitrogen from anatomic dead space. *Phase II:* mixture of gas from anatomic dead space and alveoli. *Phase III:* ("alveolar plateau") gas from alveoli. A steep slope of phase III indicates nonuniform distribution of alveolar gas. *Phase IV:* closing volume. Takeoff point of phase IV denotes beginning of airway closure in dependent portions of the lung.

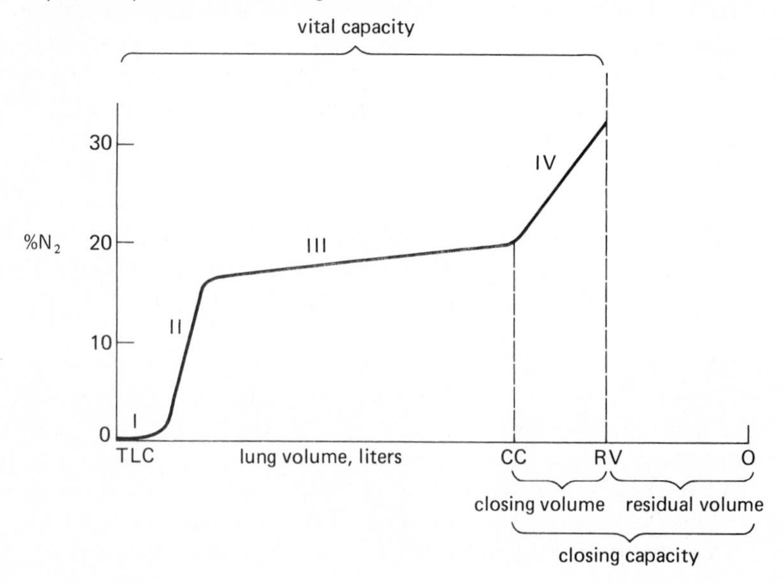

Note that in a healthy person the slope of phase III is nearly horizontal. In a patient with certain types of airways-resistance maldistribution, the phase III slope rises rapidly. This is because if certain alveoli are supplied by high-resistance airways, they fill more slowly than the normal airways during the 100% oxygen inspiration and thus have a relatively higher nitrogen concentration. During expiration they empty more slowly, and as they do, the expired nitrogen concentration rises.

As the expiration to the residual volume continues, the positive pleural surface pressure causes dynamic compression and ultimately airway closure. Because of the intrapleural pressure gradient from the upper parts of the lung to the lower parts of the lung, the airway closure first occurs in lower regions of the lung where the nitrogen concentration is the lowest. Thus, as airway closure begins, the expired nitrogen concentration rises abruptly because more and more of the expired gas is coming from alveoli in upper regions of the lung. These alveoli have the highest nitrogen concentration. The point at which the expired nitrogen concentration trace rises abruptly is the volume at which airway closure in dependent parts of the lung begins. At this point, the subject is at his or her *closing capacity,* which is equal to his or her residual volume plus the volume expired between the beginning of airway closure and the residual volume. This volume is called the *closing volume.*

Chapter 4

Blood Flow to the Lung

OBJECTIVES

The student knows the structure, function, distribution, and control of the blood supply of the lung.

1 Compares and contrasts the bronchial circulation and the pulmonary circulation.
2 Describes the anatomy of the pulmonary circulation and explains its physiological consequences.
3 Compares and contrasts the pulmonary circulation and the systemic circulation.
4 Describes and explains the effects of lung volume on pulmonary vascular resistance.
5 Describes and explains the effects of elevated intravascular pressures on pulmonary vascular resistance.
6 Lists the neural and humoral factors that influence pulmonary vascular resistance.
7 Describes the effect of gravity on pulmonary blood flow.
8 Describes the interrelationships of alveolar pressure, pulmonary arterial

pressure, and pulmonary venous pressure and their effects on the regional distribution of pulmonary blood flow.

9 Predicts the effects of alterations in alveolar pressure, pulmonary arterial and venous pressure, and body position on the regional distribution of pulmonary blood flow.

10 Describes hypoxic pulmonary vasoconstriction and discusses its role in localized and widespread alveolar hypoxia.

11 Describes the causes and consequences of pulmonary edema.

The lung receives blood flow via both the bronchial circulation and the pulmonary circulation. *Bronchial blood flow* constitutes a very small portion of the output of the left ventricle and it supplies part of the tracheobronchial tree with systemic arterial blood. *Pulmonary blood flow* constitutes the entire output of the right ventricle and it supplies the lung with the mixed venous blood draining all of the tissues of the body. It is this blood that undergoes gas exchange with the alveolar air in the pulmonary capillaries. Because the right and left ventricles are arranged in series, pulmonary blood flow is approximately equal to 100 percent of the output of the left ventricle. That is, pulmonary blood flow is equal to the cardiac output—normally about 3.5 liters/min per square meter of body surface area at rest.

There is about 250 to 300 ml of blood per square meter of body surface area in the pulmonary circulation at any instant. About 60 to 70 ml/m^2 of this blood is located in the pulmonary capillaries. It takes a red blood cell about 4 to 5 s to travel through the pulmonary circulation at resting cardiac outputs; about 0.75 s of this time is spent in pulmonary capillaries. An erythrocyte passes through a number of pulmonary capillaries as it travels through the lung. Gas exchange starts to take place in smaller pulmonary arterial vessels, which are not truly capillaries by histological standards. These arterial segments and successive capillaries may be thought of as *functional pulmonary capillaries*. In most cases in this book *pulmonary capillaries* refer to functional pulmonary capillaries rather than anatomic capillaries.

About 280 billion pulmonary capillaries supply the approximately 300 million alveoli, resulting in a potential surface area for gas exchange estimated to be 60 to 100 m^2. As was shown in Fig. 1-3, the alveoli are completely enveloped in pulmonary capillaries. The capillaries are so close to each other that some researchers have described pulmonary capillary blood flow as resembling blood flowing through two parallel sheets of endothelium held together by occasional connective tissue supports.

THE BRONCHIAL CIRCULATION

The bronchial arteries arise variably from the aorta. They supply arterial blood to the tracheobronchial tree and other structures of the lung down to the level of the terminal bronchioles. Portions of the tracheobronchial tree distal to the terminal bronchioles, including the respiratory bronchioles, alveolar ducts, and alveolar sacs, receive oxygen directly by diffusion from the alveolar air and nutrients from the mixed venous blood in the pulmonary circulation.

The blood flow in the bronchial circulation constitutes about 2 percent of the output of the left ventricle. Blood pressure in the bronchial arteries is the same as that in the other systemic arteries (disregarding differences due to hydrostatic effects, as will be discussed later in this chapter). This is much higher than the blood pressure in the pulmonary arteries, as can be seen in Fig. 4-1. The reasons for this difference will be discussed in the next section.

The venous drainage of the bronchial circulation is highly varied. Although some of the bronchial venous blood enters the azygos vein, a substantial portion of bronchial venous blood enters the *pulmonary* veins. Because the blood in the pulmonary veins has undergone gas exchange with the alveolar air (that is, the pulmonary veins contain "arterial" blood), the bronchial venous blood entering the pulmonary venous blood

Figure 4-1 Pressures [in mmHg (or torr)] in the systemic and pulmonary circulations.

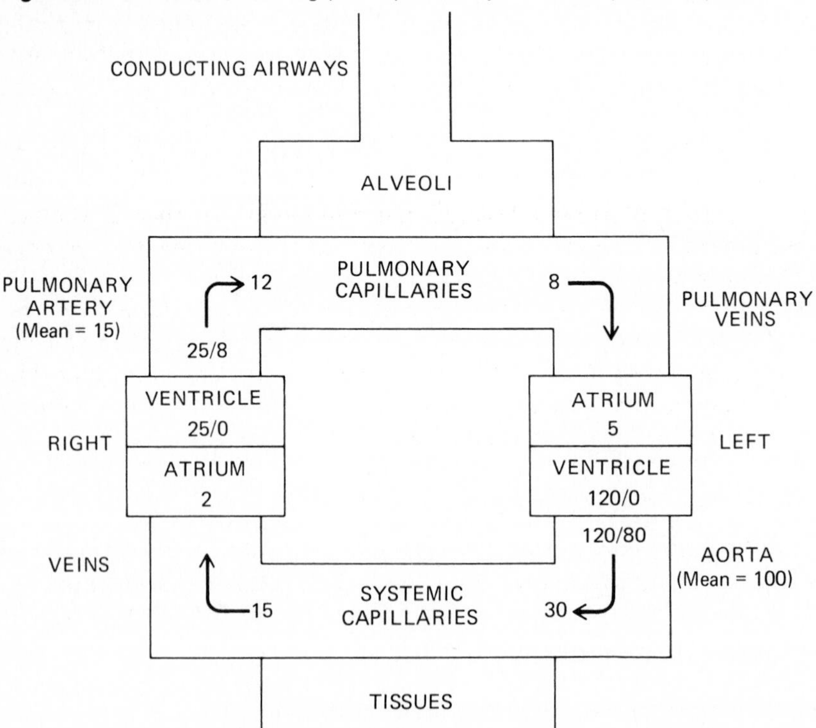

is part of the normal anatomic right-to-left shunt, which will be discussed in Chap. 5. Histologists have also identified anastomoses, or connections, between some bronchial capillaries and pulmonary capillaries and between bronchial arteries and branches of the pulmonary artery. These connections probably play little role in a normal healthy person but may open in pathological states, such as when either bronchial or pulmonary blood flow to a portion of lung is occluded. For example, if pulmonary blood flow to an area of the lung is blocked by a pulmonary embolus, bronchial blood flow to that area increases.

THE FUNCTIONAL ANATOMY OF THE PULMONARY CIRCULATION

The pulmonary circulation, from beginning to end, is much thinner-walled than corresponding parts of the systemic circulation. This is particularly true of the main pulmonary artery and its branches. The pulmonary artery rapidly subdivides into terminal branches that have thinner walls and greater internal diameters than corresponding branches of the systemic arterial tree. There is much less vascular smooth muscle in the walls of the vessels of the pulmonary arterial tree. There are no highly muscular vessels that correspond to the systemic arterioles. The pulmonary arterial tree rapidly subdivides over a short distance, ultimately branching into the approximately 280 billion pulmonary capillaries, where gas exchange occurs.

The thin walls and small amount of smooth muscle found in the pulmonary arteries have important physiological consequences. The pulmonary vessels offer much less resistance to blood flow than do the systemic arterial vessels. They are also much more distensible and compressible than systemic arterial vessels. These factors lead to much lower intravascular pressures than those found in the systemic arteries. Because the pulmonary vessels are located in the thorax and subject to alveolar and intrapleural pressures (which can change greatly), factors other than the tone of the pulmonary vascular smooth muscle may have profound effects on the pulmonary vascular resistance.

Determination of Pulmonary Vascular Resistance

Pulmonary vascular resistance cannot be measured directly but must be calculated. Poiseiulle's law, which states that for a Newtonian fluid flowing steadily through a nondistensible tube, $P_1 - P_2 = \dot{Q} \times R$, is usually used for this. This can be rearranged to

$$R = \frac{P_1 - P_2}{\dot{Q}}$$

where P_1 = pressure at the beginning of the tube (in millimeters of mercury)
P_2 = pressure at the end of the tube
$\dot{Q}$ = flow (in milliliters per minute)
R = Resistance (in millimeters of mercury per milliliter per minute)

For the pulmonary circulation, then

$$PVR = \frac{MPAP - MLAP}{PBF}$$

That is, the pulmonary vascular resistance (PVR) is equal to the mean pulmonary artery pressure (MPAP) minus the mean left atrial pressure (MLAP), with the result divided by pulmonary blood flow (PBF), which is equal to the cardiac output.

This formula, however, is only an *approximation* because blood is not a Newtonian fluid, because pulmonary blood flow is *pulsatile* (and may also be turbulent), because the pulmonary circulation is distensible (and *compressible*), and because the pulmonary circulation is a very complex branching structure. (Remember that resistances in series add directly; resistances in parallel add as reciprocals.)

As can be seen in Fig. 4-1, the intravascular pressures in the pulmonary circulation are lower than those in the systemic circulation. This is particularly striking with respect to the arterial pressures of the two circuits.

Since the right and left circulations are in series, the outputs of the right and left ventricles must be approximately equal to each other over the long run. (If they are not, blood and fluid will build up in the lungs or periphery.) If the two outputs are the same and the measured pressure drop across the systemic circulation and the pulmonary circulation are about 98 mmHg and 10 mmHg, respectively, as shown in the figure, then the pulmonary vascular resistance (PVR) must be about one-tenth that of the systemic vascular resistance (SVR), which is sometimes called TPR for *total peripheral resistance*. This low resistance to blood flow offered by the pulmonary circulation is due to the *structural* aspects of the pulmonary circulation already discussed. The pulmonary vasculature is thinner-walled, has much less vascular smooth muscle, and is generally more distensible than the systemic circulation.

Distribution of Pulmonary Vascular Resistance

The distribution of pulmonary vascular resistance can be seen by looking at the pressure drop across each of the three major components of the pulmonary vasculature: the pulmonary arteries, the pulmonary capillaries, and the pulmonary veins. As can be seen in Fig. 4-1, the resistance is fairly evenly distributed among the three components. About one-third of the resistance to blood flow is located in the pulmonary arteries, about one-third is located in the pulmonary capillaries, and about one-third is located in the pulmonary veins. This is in contrast to the systemic circulation, in which about 70 percent of the resistance to blood flow is located in the systemic arteries. Most of this resistance is in the highly muscular systemic arterioles, often called the "resistance vessels."

Consequences of Differences in Pressure Between the Systemic and Pulmonary Circulation

The pressure at the bottom of a column of a liquid is proportional to the height of the column times the density of the fluid times gravity. Thus, when the normal mean systemic arterial blood pressure is stated to be about 100 mmHg (torr), it means that the pressure developed in the aorta is equivalent to the pressure at the bottom of a column of mercury that is 100 mm high (or that it will push a column of mercury up 100 mm). Mercury is chosen for pressure measurement when high pressures are expected because it is a very dense liquid. Water is used when lower pressures are to be measured because mercury is 13.6 times as dense as water. Therefore, lower pressures, such as alveolar or pleural pressures, are expressed in centimeters of water.

Nevertheless, when mean arterial blood pressure is stated to be 100 mmHg, this is specifically with reference to the level of the left atrium. Blood pressure in the *feet* of a person who is standing is much higher than 100 mmHg because of the additional pressure exerted by the "column" of blood from the heart to the feet. In fact, blood pressure in the feet of a standing person of average height with a mean arterial blood pressure of 100 is likely to be about 180 mmHg. Since venous pressure is similarly increased in the feet (about 80 mmHg), the pressure *difference* between arteries and veins is unaffected. Conversely, pressure decreases with distance above the heart (above with respect to gravity), so that blood pressure at the top of the head may only be 40 to 50 mmHg.

The left ventricle, then, must maintain a relatively high mean arterial pressure because such high pressures are necessary to overcome hydrostatic forces and pump blood "uphill" to the brain. The apices of the lungs are a much shorter distance above the right ventricle, and so such high pressures are unnecessary.

A second consequence of the high arterial pressure in the systemic circulation is that it allows the redistribution of the left ventricular output and control of blood flow to the various tissues perfused by the cardiac output. Because the left ventricle is supplying all of the tissues of the body with blood, it must be able to meet varying demands for blood flow in different tissues under various circumstances. For example, during exercise the blood vessels supplying the exercising muscle dilate in response to the increased local metabolic demand. During severe exercise, in which blood flow to skeletal muscles increases greatly because of this mechanism (and blood flow to the skin also increases to aid in thermoregulation), blood pressure is maintained by increasing the resistance to blood flow in other vascular beds. A high pressure head is thus necessary to allow such redistributions by altering the vascular resistance to blood flow in different organs and such redistributions help to maintain the pressure head. In the pulmonary circulation such redistributions are usually unnecessary because all functioning alveolar-capillary units that are participating in gas exchange are performing the same function. The pressure head is low and the small amount of smooth muscle in the pulmonary vessels (which is in large part *responsible* for the low pressure head) makes such local redistributions unlikely. An exception to this will be seen in the section on the hypoxic pulmonary vasoconstriction.

A final consequence of the pressure difference between the systemic and pulmonary circulations is that the work load of the left ventricle (stroke work equals stroke volume times arterial pressure) is much greater than that of the right ventricle. The metabolic demand of the left ventricle is also much greater than that of the right ventricle. The difference in wall thickness of the left and right ventricles of the adult are a reminder of the much greater work load of the left ventricle.

PULMONARY VASCULAR RESISTANCE

The relatively small amounts of vascular smooth muscle, low intravascular pressures, and high distensibility of the pulmonary circulation lead to a much greater importance of extravascular effects (''passive factors'') on pulmonary vascular resistance. Gravity, body position, lung volume, alveolar and intrapleural pressure, intravascular pressures, and right ventricular output all can have profound effects on pulmonary vascular resistance without any alteration in the tone of the pulmonary vascular smooth muscle.

The Concept of a Transmural Pressure Gradient

As noted in our discussion of airways resistance in Chap. 2, for distensible-compressible vessels, the *transmural pressure gradient* is an important determinant of the vessel diameter. As the transmural pressure gradient (which is equal to pressure inside minus pressure outside) increases, the vessel diameter increases and resistance falls; as the transmural pressure decreases, the vessel is compressed and the resistance increases. *Negative* transmural pressure gradients may lead to the collapse of the vessel.

Lung Volume and Pulmonary Vascular Resistance

The effect of changes in lung volume on pulmonary vascular resistance is somewhat complex. As shown in Fig. 4-2, two different groups of pulmonary vessels must be taken into consideration: those exposed to alveolar pressure and those outside the alveoli. These two groups are referred to as the *alveolar* and *extraalveolar* vessels, respectively.

As lung volume increases during a normal negative pressure inspiration, the alveoli increase in volume. As they expand, the vessels found between them, mainly pulmonary capillaries, are compressed. At high lung volumes, then, the resistance to blood flow offered by the alveolar vessels increases; at low lung volumes, the resistance to blood flow offered by the alveolar vessels decreases. This can be seen in the *"alveolar"* curve in Fig. 4-3.

One group of the *extraalveolar* vessels, the larger arteries and veins, is exposed to the intrapleural pressure. As lung volume is increased by making the intrapleural pressure more negative, their transmural pressure gradient increases and they distend. Another factor tending to decrease the resistance to blood flow offered by the extraalveolar vessels at higher lung volumes is radial traction by the connective tissue and alveolar septa holding the larger vessels in place in the lung. Thus, at high lung volumes (attained by normal negative-pressure breathing), the resistance to blood flow offered by the extraalveolar vessels decreases, as shown in Fig. 4-3.

Figure 4-2 Schematic illustration of "alveolar" and "extraalveolar" pulmonary vessels during an inspiration. The alveolar vessels, pulmonary capillaries, are exposed to the expanding alveoli and are compressed. The extraalveolar vessels, here shown exposed to the intrapleural pressure, expand as the intrapleural pressure becomes more negative during the inspiration.

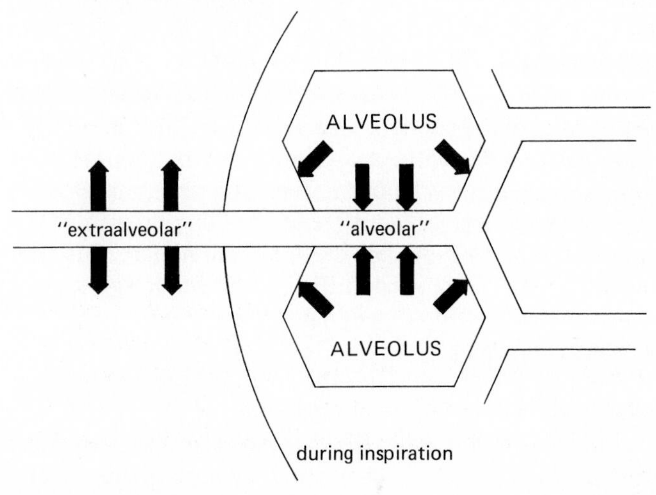

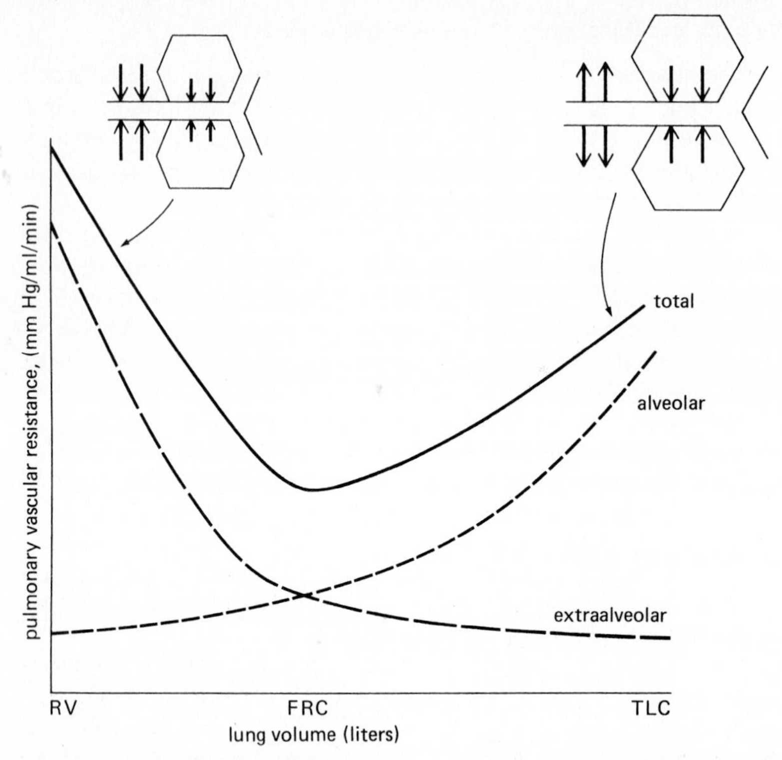

Figure 4-3 The effects of lung volume on pulmonary vascular resistance. PVR is lowest near the FRC and increases at both high and low lung volumes because of the combined effects on the alveolar and extraalveolar vessels. To achieve low lung volumes, one must generate *positive* intrapleural pressures and the extraalveolar vessels are compressed, as seen at left in the figure. (*Graph after Murray, 1976. Reproduced with permission.*)

During a forced expiration to low lung volumes, however, intrapleural pressure becomes very positive. The resistance to blood flow offered by the extraalveolar vessels increases greatly, as seen at left in Fig. 4-3.

Because the alveolar and extraalveolar vessels may be thought of as two groups of resistances in series with each other, the resistances of the alveolar and extraalveolar vessels are additive at any lung volume. Thus the effect of changes in lung volume on the *total* pulmonary vascular resistance gives the U-shaped curve seen in Fig. 4-3. Pulmonary vascular resistance is lowest near the functional residual capacity and increases at both high and low lung volumes.

A second type of extraalveolar vessel is the so-called corner vessel or extraalveolar capillary shown in Fig. 4-4. Although these vessels are found between alveoli, their locations at junctions of alveolar septa give them different mechanical properties, as shown in the figure. Expansion of the alveoli during inspiration increases the wall tension of the alveolar

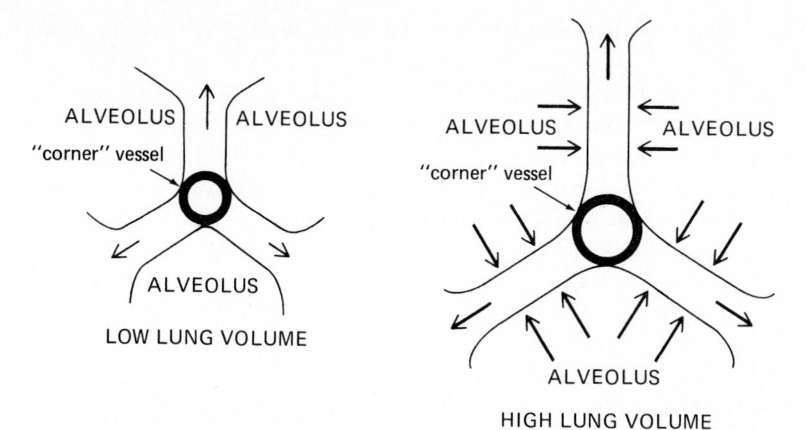

Figure 4-4 Schematic illustration of the extraalveolar "corner vessels," located at junctions of alveolar septa. Expansion of the alveoli causes radial traction on the corner vessels and expands them. The alveolar vessels are compressed at high lung volumes.

septa and the corner vessels or extraalveolar capillaries are distended, whereas the alveolar capillaries are compressed.

Also note that during *mechanical* positive-pressure ventilation, intrapleural pressure is *positive* during inspiration. In this case both the alveolar and extraalveolar vessels are compressed as lung volume increases, and the resistance to blood flow offered by both alveolar and extraalveolar vessels increases during lung inflation. This is especially a problem during mechanical positive-pressure ventilation with positive end-expiratory pressure (PEEP). In this situation, intrapleural pressure is positive during both inspiration and expiration. Pulmonary vascular resistance is elevated in both alveolar and extraalveolar vessels throughout the respiratory cycle. In addition, because intrapleural pressure is continuously positive, the other intrathoracic blood vessels are also subjected to decreased transmural pressure gradients and the venae cavae, which have low intravascular pressure, are also compressed. If cardiovascular reflexes are unable to adjust to this situation, cardiac output may fall precipitously because of decreased venous return and high pulmonary vascular resistance.

Recruitment and Distensibility

During exercise, cardiac output can increase severalfold without a correspondingly great increase in mean pulmonary artery pressure. Although the mean pulmonary artery pressure does increase, the increase is only a few millimeters of mercury, even if cardiac output has doubled or tripled. Since the pressure drop across the pulmonary circulation is proportional to the cardiac output times the pulmonary vascular resistance (that is,

$\Delta P = \dot{Q} \times R$), this must indicate a decrease in pulmonary vascular resistance.

This fall in pulmonary vascular resistance appears to be *passive*—that is, it is not due to changes in the tone of pulmonary vascular smooth muscle caused by neural mechanisms or humoral agents. In fact, a fall in pulmonary vascular resistance in response to increased blood flow or even an increase in perfusion pressure can be demonstrated in a vascularly isolated perfused lung, as was used to obtain the data summarized in Fig. 4-5.

In this experiment the blood vessels of the left lung of a dog were isolated, cannulated, and perfused with a pump. The lung was ventilated with a mechanical respirator. Mean pulmonary artery pressure was increased progressively by increasing the pump output. As can be seen from the graph, increasing the mean pulmonary artery pressure by increasing the blood flow to the lung caused a decrease in the calculated pulmonary

Figure 4-5 The effect of mean pulmonary artery pressure on pulmonary vascular resistance. Increased mean pulmonary artery pressure decreases pulmonary vascular resistance. (*After Borst, 1956. Reproduced by permission of the American Heart Association, Inc.*)

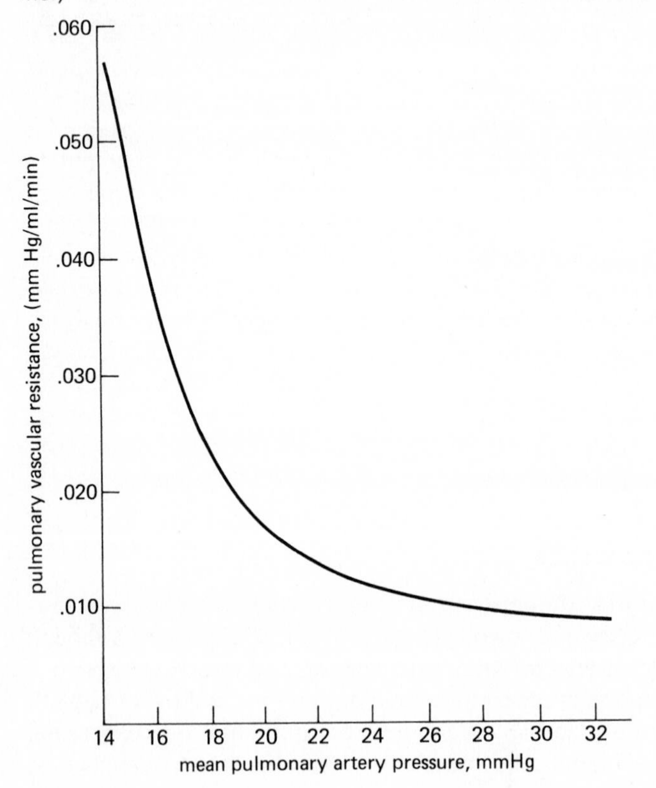

vascular resistance. Increasing the left atrial pressure also decreased pulmonary vascular resistance in these studies.

There are two different mechanisms that can explain this decrease in pulmonary vascular resistance in response to elevated blood flow and perfusion pressure. These two mechanisms, *recruitment* and *distention,* are illustrated in Fig. 4-6.

Recruitment As indicated in the diagram, at resting cardiac outputs, not all of the pulmonary capillaries are perfused. A substantial number of capillaries is likely unperfused because of hydrostatic effects that will be discussed later in this chapter. Others may be unperfused because they have a relatively high *critical opening pressure.* That is, these vessels, because of their high vascular smooth muscle tone, require a higher perfusion pressure than that solely necessary to overcome hydrostatic forces. It is not likely that the critical opening pressures for pulmonary blood vessels are very great, since they have so little smooth muscle.

Figure 4-6 Schematic illustration of the mechanisms by which increased mean pulmonary artery pressure may decrease pulmonary vascular resistance. The upper figure shows a group of pulmonary capillaries, some of which are not perfused. At left, the previously unperfused capillaries are recruited (opened) by the increased perfusion pressure. At right, the increased perfusion pressure has distended those vessels already open.

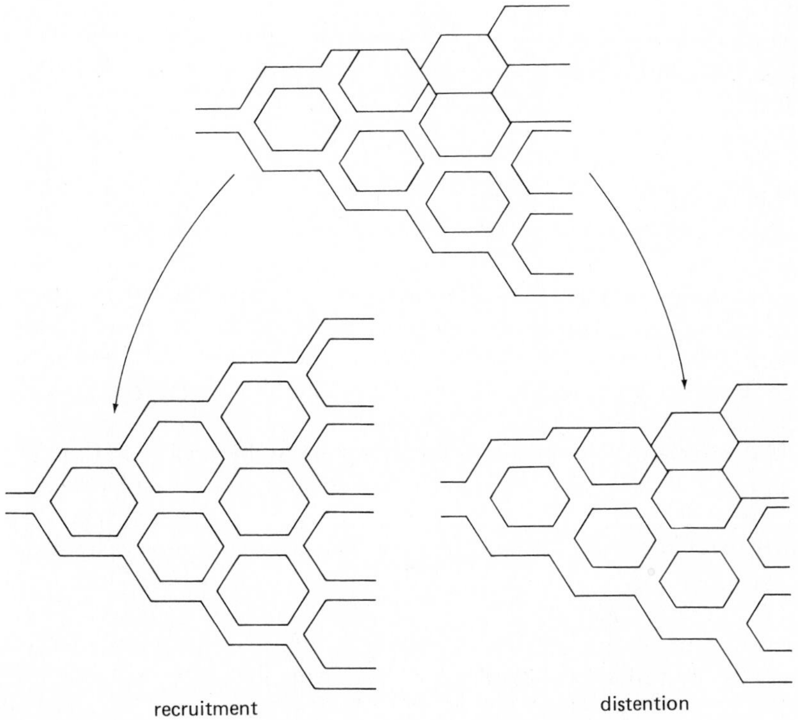

Increasing blood flow increases the mean pulmonary artery pressure, which opposes hydrostatic forces and exceeds the critical opening pressure in vessels with high vascular tone. This series of events opens new parallel pathways for blood flow, which lowers the pulmonary vascular resistance. This opening of new pathways is called *recruitment.* Note that *decreasing* the cardiac output or pulmonary artery pressure can result in a *derecruitment* of pulmonary capillaries.

Distention The distensibility of the pulmonary vasculature has already been discussed in this chapter. As perfusion pressure increases, the transmural pressure gradient of the pulmonary blood vessels increases, causing distention of the vessels. This increases their radii and decreases their resistance to blood flow.

Recruitment or Distention? The answer to the question of whether it is recruitment or distention that causes the decreased pulmonary vascular resistance seen with elevated perfusion pressure is probably *both. Recruitment* of pulmonary capillaries probably occurs at low pulmonary vascular pressures and *distention* at higher pressures.

Control of Pulmonary Vascular Smooth Muscle

Pulmonary vascular smooth muscle is responsive to both *neural* and *humoral* influences. These produce "active" alterations in pulmonary vascular resistance, as opposed to those "passive" factors discussed in the previous section. A final passive factor, *gravity,* will be discussed later in this chapter.

Neural Effects The pulmonary vasculature is innervated by both sympathetic and parasympathetic fibers of the autonomic nervous system. The innervation of pulmonary vessels is relatively sparse in comparison to that of systemic vessels. There is relatively more innervation of the larger, elastic vessels and less of the smaller, more muscular vessels. There appears to be no innervation of vessels smaller than 30 μm in diameter. There does not appear to be much innervation of intrapulmonary veins and venules.

The effects of stimulation of the sympathetic innervation of the pulmonary vasculature are somewhat controversial. Some investigators have demonstrated an increase in pulmonary vascular resistance with sympathetic stimulation of the innervation of the pulmonary vasculature, whereas others showed a decreased *distensibility* with no change in calculated pulmonary vascular resistance. Stimulation of the parasympathetic innervation of the pulmonary vessels generally causes vasodilation, although its physiological function is not known.

Humoral Effects The catecholamines epinephrine and norepinephrine both increase pulmonary vascular resistance when injected into the pulmonary circulation. Serotonin and histamine, found in the lung in mast cells, are both pulmonary vasoconstrictors. Certain prostaglandins, such as $PGF_{2\alpha}$ and PGE_2, are also pulmonary vasoconstrictors, as are some prostaglandin precursors and breakdown products. Alveolar hypoxia and hypercapnia also cause pulmonary vasoconstriction, as will be discussed later in this chapter. Acetylcholine, the beta-adrenergic agonist isoproterenol, and certain prostaglandins, such as PGE_1, are all pulmonary vasodilators.

THE REGIONAL DISTRIBUTION OF PULMONARY BLOOD FLOW: THE ZONES OF THE LUNG

Determinations of the regional distribution of pulmonary blood flow have shown that *gravity* is another important "passive" factor affecting local pulmonary vascular resistance and the relative perfusion of different regions of the lung. The interaction of the effects of gravity and extravascular pressures may have a profound influence on the relative perfusion of different areas of the lung.

Measurement of Total Pulmonary Blood Flow

The total pulmonary blood flow (which is the cardiac output) can currently be determined clinically in several ways, each of which has the disadvantage of being *invasive* to the patient (that is, requiring minor surgery).

The Fick Principle In 1870, Adolf Fick pointed out the following relationship, now known as the Fick principle: The total amount of oxygen absorbed by the body per minute ($\dot{V}O_2$) must be equal to the cardiac output in milliliters per minute ($\dot{Q}t$) times the difference in oxygen *content,* in ml O_2 per 100 ml blood, between the arterial and mixed venous blood ($Ca_{O_2} - C\bar{v}_{O_2}$).

$$\dot{V}O_2 = \dot{Q}t \times (Ca_{O_2} - C\bar{v}_{O_2})$$

$$\dot{Q}t = \frac{\dot{V}O_2}{Ca_{O_2} - C\bar{v}_{O_2}}$$

Both arterial and mixed venous (which is equal to pulmonary artery) blood must be sampled in this method.

The Indicator Dilution Technique In this method, a known amount of an indicator dye, such as indocyanine green (which stays in the blood vessels), is injected intravenously as a bolus. The systemic arterial dye concentration is monitored continuously with a densitometer as the dye passes through the aorta. Correction must be made for *recirculating* dye because the concentration change of interest is that in a single pass through the pulmonary circulation. A curve of the dye concentration as it changes with time is constructed and then the area under the curve is determined by integration. Dividing this amount by the time of passage of the dye gives the average concentration of dye through the passage. If the cardiac output is high, the dye concentration falls rapidly, and so the area under the curve is small and the average dye concentration is low. If the cardiac output is low, the area under the curve is large and the average dye concentration is high. The cardiac output ($\dot{Q}t$) is equal to the amount of dye injected in milligrams (I) divided by the mean dye concentration in milligrams per milliliter (c) times the time of passage (t) in seconds:

$$\dot{Q}t = \frac{I}{ct}$$

The calculations are usually done automatically in newer densitometers. Both arterial and venous catheters are necessary for this technique.

The Thermal Dilution Technique This method is similar in principle to the indicator dilution technique. Cold fluid, for example, saline, is injected into a central vein and the change in temperature of the blood downstream is monitored continuously with a thermistor. With high cardiac outputs, the temperature rises rapidly; with low cardiac outputs the temperature rises slowly. The advantage of this method is that the insertion of a single intravenous catheter is the only necessary surgical procedure. A type of catheter known as a quadruple-lumen Swan-Ganz catheter is used. One lumen is connected to a tiny inflatable balloon at the end of the catheter. During the insertion of the catheter, the balloon is inflated so that the tip of the catheter "floats" in the direction of blood flow: through the right atrium and ventricle and into the pulmonary artery. The balloon is then deflated. A second lumen carries the thermistor wire to the end of the catheter. A third lumen travels only part of the way down the catheter so it opens into a central vein. This lumen is used for the injection of the cold solution. The final lumen, at the end of the catheter, is open to the pulmonary artery and it allows pulmonary artery pressure to be monitored. This monitoring is necessary because the only way the physician knows that the catheter is placed properly is by recognizing the characteristic pulmonary artery pressure trace (unless a fluoroscope is used).

The temperature change after the injection is monitored by a cardiac output "computer" that automatically calculates the cardiac output from the temperature of the injected substance, the original blood temperature, and the temperature change of the blood with time.

Determination of Regional Pulmonary Blood Flow

Regional pulmonary blood flow can be determined by pulmonary angiography, by lung scans after injection of ^{131}I–labeled macroaggregates of albumin, and by lung scans after the infusion of dissolved radio-labeled gases such as ^{133}Xe.

Pulmonary Angiography A radioopaque substance is injected into the pulmonary artery and its movement through the lung is monitored during fluoroscopy. Unperfused areas secondary to vascular obstruction by emboli or from other causes are evident because none of the radioopaque substance enters these areas.

Macroaggregates of Albumin Radio-labeled macroaggregates of albumin in the size range of 10 to 50 μm are injected in small quantities into a peripheral vein. Most of these become trapped in small pulmonary vessels as they enter the lung. Lung scans for radioactivity demonstrate the perfused areas of the lung. The aggregates fragment and are removed from the lung within a day or so.

^{133}Xe ^{133}Xe is dissolved in saline and injected intravenously. Xenon is not particularly soluble in saline or blood, and so it comes out of solution in the lung and enters the alveoli. If the ^{133}Xe is well mixed in the blood as it enters the pulmonary artery, then the amount of radioactivity coming into a region of the lung is proportional to the amount of blood flow to that area. By making corrections for regional lung volume, the blood flow *per unit volume* of a region of the lung can be determined.

The Regional Distribution of Pulmonary Blood Flow

If ^{133}Xe is used to determine regional pulmonary blood flow in a person seated upright or standing up, a pattern like that shown in Fig. 4-7 is seen. There is greater blood flow per unit volume (''per alveolus'') to lower regions of the lung than to upper regions of the lung. Note that the test was made with the subject at the total lung capacity.

If the subject lies down, this pattern of regional perfusion is altered so that perfusion to the *anatomically* upper and lower portions of the lung is roughly evenly distributed, but blood flow per unit volume is still greater in the more gravity-dependent regions of the lung. For example, if the subject were to lie down on his left side, the left lung would receive more blood flow per unit volume than would the right lung. Exercise, which increases the cardiac output, increases the blood flow per unit volume to all regions of the lung, but the perfusion gradient persists so that there is

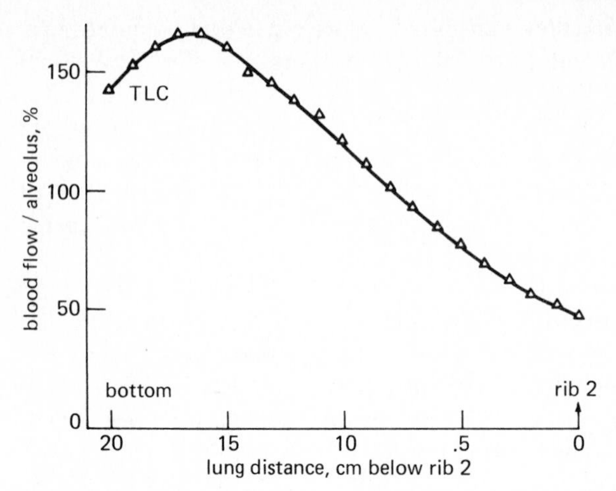

Figure 4-7 Relative blood flow per alveolus (100% = perfusion of each alveolus if all were perfused equally) versus distance from the bottom of the lung in a human seated upright. Measurement of regional blood flow was determined using an intravenous injection of ^{133}Xe. (*Redrawn from Hughes, 1968, with permission.*)

still relatively greater blood flow per unit volume in more gravity-dependent regions of the lung.

The reason for this gradient of regional perfusion of the lung is obviously *gravity*. As already discussed, the pressure at the bottom of a column of a liquid is proportional to the height of the column times the density of the liquid times gravity. Thus, the intravascular pressures in more gravity-dependent portions of the lung are greater than those in upper regions. Because the pressures are greater in the more gravity-dependent regions of the lung, the *resistance to blood flow is lower* in lower regions of the lung because of *recruitment* or *distention* of vessels in these regions. It is therefore not only gravity but also the peculiar characteristics of the pulmonary circulation that cause the increased blood flow to more dependent regions of the lung. After all, the same hydrostatic effects occur to an even greater extent in the left side of the circulation, but the thick walls of the systemic arteries are not affected.

The Interaction of Gravity and Extravascular Pressure: The Zones of the Lung

Experiments done on excised, perfused, upright animal lungs have demonstrated the same gradient of increased perfusion per unit volume from the top of the lung to the bottom. When the experiments were done at low pump outputs, so that the pulmonary artery pressure was low, the uppermost regions of the lung received no blood flow. Perfusion of the lung ceased at the point at which alveolar pressure (PA) was just equal to pulmonary arterial pressure (Pa). Above this point, there was no perfusion because the transmural pressure across capillary walls was negative. Below this point, perfusion per unit volume increased steadily with increased distance down the lung.

Thus, under circumstances in which alveolar pressure is higher than pulmonary artery pressure in the upper parts of the lung, no blood flow occurs in that region, and the region is referred to as being in zone 1, as shown in Fig. 4-8. Any zone 1, then, is ventilated but not perfused. It is *alveolar dead space*. Fortunately, during normal, quiet breathing in a person with a normal cardiac output, pulmonary artery pressure, even in the uppermost regions of the lung, is higher than alveolar pressure, so there is no zone 1.

The lower portion of the lung in Fig. 4-8 is said to be in zone 3. In this region, the pulmonary artery pressure and the pulmonary vein pressure (Pv) are both higher than alveolar pressure. The driving pressure for blood flow through the lung in this region is simply pulmonary artery pressure minus pulmonary vein pressure. Note that this driving pressure stays constant as one moves further down the lung in zone 3 because the hydrostatic pressure effects are the same for both the arteries and the veins.

The middle portion of the lung in Fig. 4-8 is in zone 2. In zone 2, pulmonary artery pressure is greater than alveolar pressure, and so blood flow does occur. Nevertheless, because alveolar pressure is greater than pulmonary vein pressure, the *effective* driving pressure for blood flow is pulmonary artery pressure minus alveolar pressure in zone 2. Notice that in zone 2 (at right in Fig. 4-8) the increase in blood flow per distance down the lung is greater than it is in zone 3. This is because the upstream driving pressure, the pulmonary artery pressure, increases according to the hy-

Figure 4-8 The zones of the lung. The effects of gravity and alveolar pressure on the perfusion of the lung. Described in text. (*Redrawn from West, 1964, with permission.*)

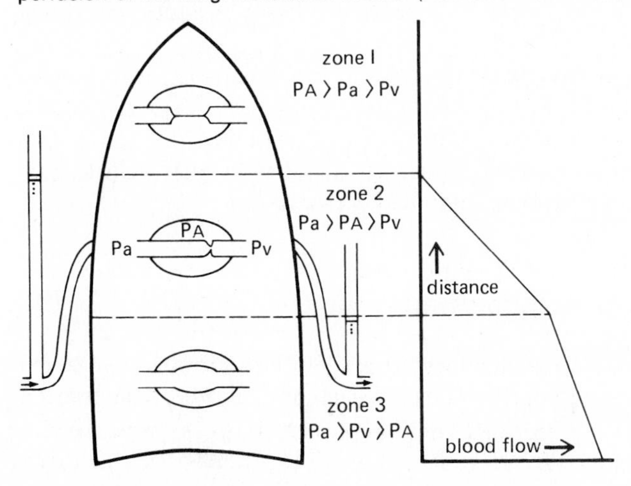

drostatic pressure increase, but the effective downstream pressure, alveolar pressure, is constant throughout the lung.

To summarize then: In *zone 1*,

$$P_A > Pa > Pv$$

and there is no blood flow (except perhaps in "corner vessels," such as those in Fig. 4-4, which are not exposed to alveolar pressure); in *zone 2*,

$$Pa > P_A > Pv$$

and the effective driving pressure for blood flow is $Pa - P_A$; in *zone 3*,

$$Pa > Pv > P_A$$

and the driving pressure for blood flow is $Pa - Pv$.

A most important fact to realize is that the boundaries between the zones are dependent on physiological conditions—they are not fixed anatomic landmarks. Alveolar pressure changes during the course of each breath. During eupneic breathing these changes are only a few centimeters of H_2O, but they may be much greater during speech, exercise, or other conditions. A patient on a positive-pressure ventilator with positive end-expiratory pressure may have substantial amounts of zone 1 because alveolar pressure is always high. Similarly, after a hemorrhage or during general anesthesia pulmonary blood flow and pulmonary artery pressure are low and zone 1 conditions are also likely. During exercise, cardiac output and pulmonary artery pressure increase and any existing zone 1 will be recruited to zone 2. The boundary between zones 2 and 3 will move upward as well. Pulmonary artery pressure is highly pulsatile so the borders between the zones probably even move up a bit with each contraction of the right ventricle.

Changes in lung volume also affect the regional distribution of pulmonary blood flow and will therefore affect the boundaries between zones. Finally, changes in body position alter the orientation of the zones with respect to the anatomic locations in the lung, but the same relationships exist with respect to gravity and alveolar pressure.

HYPOXIC PULMONARY VASOCONSTRICTION

Alveolar hypoxia or atelectasis causes an active vasoconstriction in the pulmonary circulation. The site of vascular smooth muscle constriction appears to be in the arterial (precapillary) vessels very close to the alveoli.

Mechanism of Hypoxic Pulmonary Vasoconstriction

The mechanism of hypoxic pulmonary vasoconstriction is not completely understood. The response occurs locally, that is, only in the area of the alveolar hypoxia. Connections to the central nervous system are not necessary: An isolated, excised lung, perfused with blood by a mechanical pump with a constant output, exhibits an increased perfusion pressure when ventilated with hypoxic gas mixtures. This situation indicates that the increase in pulmonary vascular resistance can occur without the influence of extrinsic nerves. Hypoxia may act directly on the vascular smooth muscle or may cause the release of a vasoactive substance from the pulmonary parenchyma or mast cells in the area. Histamine, serotonin, catecholamines, and prostaglandins have all been suggested as the mediator substance, but none appears to completely mimic the response. Possibly several mediators may act together.

The response is graded—constriction begins to occur at alveolar P_{O_2}'s in the range 100 to 150 torr and increases until $P_{A_{O_2}}$ falls to about 20 to 30 torr.

Physiological Function of Hypoxic Pulmonary Vasoconstriction

The function of hypoxic pulmonary vasoconstriction in localized hypoxia is fairly obvious. If an area of the lung becomes hypoxic because of airway obstruction or if localized atelectasis occurs, any mixed venous blood flowing to that area will undergo little or no gas exchange and will mix with blood draining well-ventilated areas of the lung as it enters the left atrium. This mixing will lower the overall arterial P_{O_2} and may even increase the arterial P_{CO_2}. The hypoxic pulmonary vasoconstriction diverts mixed venous blood flow away from poorly ventilated areas of the lung by locally increasing the vascular resistance and therefore sends it to better ventilated areas of the lung.

In hypoxia of the whole lung, such as might be encountered at high altitude or in hypoventilation, hypoxic pulmonary vasoconstriction occurs throughout the lung. Even this may be useful in increasing gas exchange because greatly increasing the pulmonary artery pressure recruits many previously unperfused pulmonary capillaries, increasing the surface area available for gas diffusion, as will be discussed in Chap. 6, and improving the matching of ventilation and perfusion, as will be discussed in the next chapter.

The problem with hypoxic pulmonary vasoconstriction is that it is not a very strong response, which is expected because there is so little smooth muscle in the pulmonary vasculature. Very high pulmonary artery pressures can interfere with hypoxic pulmonary vasoconstriction, as can other physiological disturbances, such as alkalosis.

Alveolar hypercapnia (high carbon dioxide) also causes pulmonary vasoconstriction. It is not clear whether this occurs by the same mechanism as that of hypoxic pulmonary vasoconstriction.

PULMONARY EDEMA

Pulmonary edema is the extravascular accumulation of fluid in the lung. This pathological condition may be caused by one or more of a number of physiological abnormalities, but the result is inevitably impaired gas transfer. As the edema fluid builds up, first in the interstitium and later in alveoli, diffusion of gases, particularly oxygen (see Chap. 6), decreases.

The Factors Influencing Liquid Movement in the Pulmonary Capillaries

The Starling equation describes the movement of liquid across the capillary endothelium:

$$\dot{Q}_f = K_f(P_C - P_{is}) - \sigma(\pi_{pl} - \pi_{is})$$

where $\dot{Q}_f$ = *net* flow of fluid
K_f = capillary filtration coefficient. This describes the permeability characteristics of the membrane to fluids.
P_C = capillary hydrostatic pressure
P_{is} = hydrostatic pressure of the interstitial liquid
σ = reflection coefficient. This describes the ability of the membrane to prevent extravasation of solute particles.
π_{pl} = colloid osmotic (oncotic) pressure of the plasma
π_{is} = colloid osmotic pressure of the interstitial fluid

The equation is shown schematically in Fig. 4-9. The Starling equation is very useful in understanding the potential causes of pulmonary edema, even though only the plasma colloid osmotic pressure (π_{pl}) can be accurately measured.

Lymphatic Drainage of the Lung

Any fluid that makes its way into the pulmonary interstitium must be removed by the lymphatic drainage of the lung. The volume of lymph flow from the human lung is now believed to be as great as that from other organs under normal circumstances and is capable of increasing as much as tenfold under pathological conditions. It is only when this large safety factor is overwhelmed that pulmonary edema occurs.

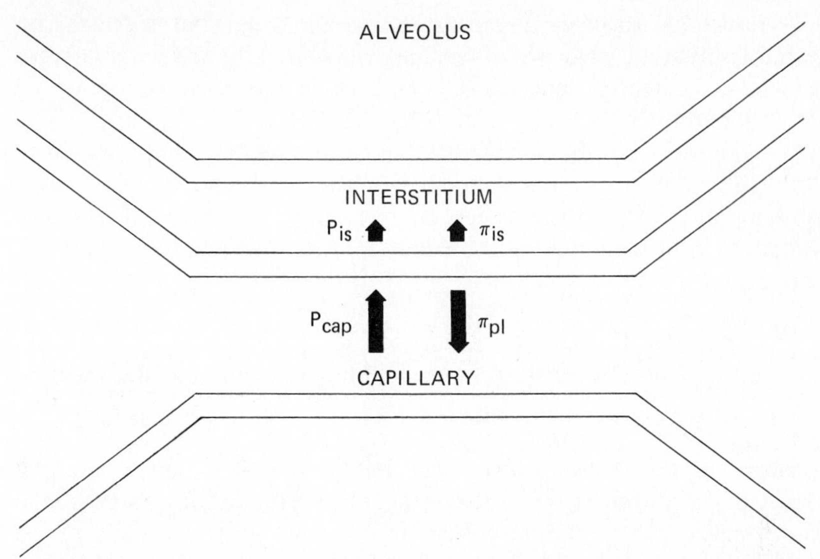

Figure 4-9 Schematic illustration of the factors affecting liquid movement from the pulmonary capillaries. P_{cap} = capillary hydrostatic pressure; P_{is} = interstitial hydrostatic pressure; π_{pl} = plasma colloid osmotic pressure; π_{is} = interstitial colloid osmotic pressure. (P_{is} is assumed to be negative.)

Conditions That May Lead to Pulmonary Edema

The Starling equation provides a useful method of categorizing most of the potential causes of pulmonary edema.

Permeability Infections, circulating or inhaled toxins, oxygen toxicity, and other factors that destroy the integrity of the capillary endothelium lead to localized or generalized pulmonary edema.

Capillary Hydrostatic Pressure The capillary hydrostatic pressure is estimated to be about 7 to 10 mmHg under normal conditions. If the capillary hydrostatic pressure increases dramatically, the filtration of fluid across the capillary endothelium will increase greatly and enough fluid may leave the capillaries to exceed the lymphatic drainage. The pulmonary capillary hydrostatic pressure often increases secondary to problems in the left side of the circulation, such as infarction of the left ventricle, left ventricular failure, or mitral stenosis. As left atrial pressure and pulmonary venous pressure rise because of accumulating blood, the pulmonary capillary hydrostatic pressure also increases. Other causes of elevated pulmonary capillary hydrostatic pressure include overzealous administration of intravenous fluids by the physician and diseases that occlude the pulmonary veins.

Interstitial Hydrostatic Pressure Some investigators believe the interstitial hydrostatic pressure of the lung to be slightly positive, whereas others have shown evidence that it may be in the range of −5 to −7 cmH_2O. Conditions that would decrease the interstitial pressure would increase the tendency for pulmonary edema to develop. These appear to be limited mainly to potential actions of the physician, such as rapid evacuation of chest fluids or reduction of pneumothorax. As should also be noted, as fluid accumulates in the interstitium, the interstitial hydrostatic pressure increases, which helps limit further fluid extravasation.

The Reflection Coefficient Any situation that permits more solute to leave the capillaries will lead to more fluid movement out of the vascular space.

Plasma Colloid Osmotic Pressure Decreases in the colloid osmotic pressure of the plasma, which helps retain fluid in the capillaries, may lead to pulmonary edema. Plasma colloid osmotic pressure, normally in the range of 25 to 28 mmHg, falls in hypoproteinemia or overadministration of intravenous solutions.

Interstitial Colloid Osmotic Pressure As noted above, increased concentration of solute in the interstitium will pull fluid from the capillaries.

Lymphatic Insufficiency Conditions that block the lymphatic drainage of the lung, such as tumors or scars, may predispose toward pulmonary edema.

Other Conditions Associated with Pulmonary Edema Pulmonary edema is often seen associated with head injury, heroin overdose, and high altitude. The cause of the edema formation in these conditions is not known, although high-altitude pulmonary edema may be caused by high pulmonary artery pressures secondary to the hypoxic pulmonary vasoconstriction.

Chapter 5

Ventilation-Perfusion Relationships and Hypoxia

OBJECTIVES

The student understands the importance of the matching of ventilation and perfusion in the lung.

1. Predicts the consequences of mismatched ventilation and perfusion.
2. Describes the methods used to assess the matching of ventilation and perfusion.
3. Describes the methods used to determine the uniformity of the distribution of the inspired gas and pulmonary blood flow.
4. Explains the regional differences in the matching of ventilation and perfusion of the normal upright lung.
5. Predicts the consequences of the regional differences in the ventilation and perfusion of the normal upright lung.
6. Classifies and explains the causes of tissue hypoxia.

Gas exchange between the alveoli and the pulmonary capillary blood occurs by diffusion, as will be discussed in the next chapter. Diffusion of oxygen and carbon dioxide occurs passively, according to their concentration gradients across the alveolar-capillary barrier. These concentration gradients must be maintained by ventilation of the alveoli and perfusion of the pulmonary capillaries.

THE CONCEPT OF MATCHING VENTILATION AND PERFUSION

Alveolar ventilation brings oxygen into the lung and removes carbon dioxide from it. Similarly, the mixed venous blood brings the carbon dioxide into the lung and takes up alveolar oxygen. The alveolar P_{O_2} and P_{CO_2} are thus determined by the relationship between alveolar ventilation and perfusion. Alterations in the ratio of ventilation to perfusion, called the *$\dot{V}_A/\dot{Q}_C$ ratio,* will result in changes in the alveolar P_{O_2} and P_{CO_2}, as well as in changes in gas delivery to or removal from the lung.

Alveolar ventilation is normally about 4 to 6 liters/min and pulmonary blood flow (which is equal to cardiac output) has a similar range, and so the ratio of ventilation to perfusion for the whole lung is in the range of 0.8 to 1.2. Ventilation and perfusion must be matched, however, on the *alveolar-capillary level,* and the $\dot{V}_A/\dot{Q}_C$ for the whole lung is really of interest only as an approximation of the situation in all the alveolar-capillary units of the lung. For instance, suppose that all 5 liters/min of the cardiac output went to the left lung and all 5 liters/min of alveolar ventilation went to the right lung. The whole lung $\dot{V}_A/\dot{Q}_C$ would be 1.0, but there could be no gas exchange because there could be no gas diffusion between the alveoli and the pulmonary capillaries.

CONSEQUENCES OF HIGH AND LOW $\dot{V}_A/\dot{Q}_C$

Figure 5-1 shows the consequences of alterations in the relationship of ventilation and perfusion for a hypothetical alveolar-capillary unit. Unit A has a normal $\dot{V}_A/\dot{Q}_C$ ratio. Inspired air enters the alveolus with a P_{O_2} of about 150 torr and a P_{CO_2} of nearly 0 torr. Mixed venous blood enters the pulmonary capillary with a P_{O_2} of about 40 torr and a P_{CO_2} of about 45 torr. This results in an alveolar P_{O_2} of about 100 torr and an alveolar P_{CO_2} of 40 torr, as was discussed in Chap. 3. The partial pressure gradient for oxygen diffusion is thus about 100 − 40, or 60, torr; the partial pressure gradient for CO_2 is about 45 − 40 torr.

The airway supplying unit B has become completely occluded. Its $\dot{V}_A/\dot{Q}_C$ is zero. As time goes on, the air trapped in the alveolus equilibrates by diffusion with the gas dissolved in the mixed venous blood entering the alveolar-capillary unit. (If the occlusion persists, the alveolus is likely to collapse.) No gas exchange can occur and any blood perfusing this alveolus will leave it exactly as it entered it. Unit B is therefore acting as a right-to-left shunt.

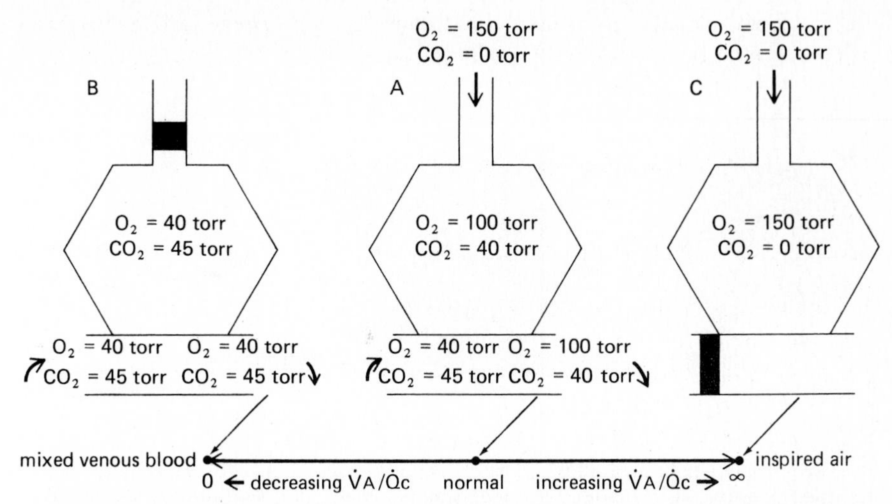

Figure 5-1 The effect of changes in the ventilation-perfusion ratio on the alveolar P_{O_2} and P_{CO_2}. *A.* Normal $\dot{V}_A/\dot{Q}_C$. *B.* $\dot{V}_A/\dot{Q}_C = 0$. *C.* $\dot{V}_A/\dot{Q}_C$ is infinite. (*Reproduced with permission from West, 1977a.*)

The blood flow to unit C is blocked by a pulmonary embolus and unit C is therefore completely unperfused. It has an infinite $\dot{V}_A/\dot{Q}_C$ ratio. Since no oxygen can diffuse from the alveolus into pulmonary capillary blood and since no carbon dioxide can enter the alveolus from the blood, the P_{O_2} of the alveolus is approximately 150 torr and its P_{CO_2} is approximately zero. That is, the gas composition of this unperfused alveolus is the same as that of inspired air. Unit C is alveolar dead space. If unit C were unperfused because its alveolar pressure exceeded its precapillary pressure (rather than because of an embolus), then it would also correspond to part of zone 1.

Units B and C represent the two extremes of a *continuum* of ventilation-perfusion ratios. The ventilation-perfusion ratio of a particular alveolar-capillary unit can fall anywhere along this continuum, as shown in Fig. 5-1. The alveolar P_{O_2} and P_{CO_2} of such units will therefore fall between the two extremes shown in the figure: Units with *low* ventilation-perfusion ratios will have relatively low P_{O_2}'s and high P_{CO_2}'s; units with *high* ventilation-perfusion ratios will have relatively high P_{O_2}'s and low P_{CO_2}'s. This is demonstrated graphically in an O_2-CO_2 diagram such as that seen in Fig. 5-2, which shows the results of mathematical calculations for alveolar P_{O_2}'s and P_{CO_2}'s for ventilation-perfusion ratios between zero (designated as $\bar{v}$, for mixed venous blood) and infinity (designated as I, for inspired air). The resulting curve is known as *the ventilation-perfusion ratio line*. This simple O_2-CO_2 diagram can be modified to include correction lines for other factors, e.g., the respiratory exchange ratios of the alveoli and the blood or the dead space. The posi-

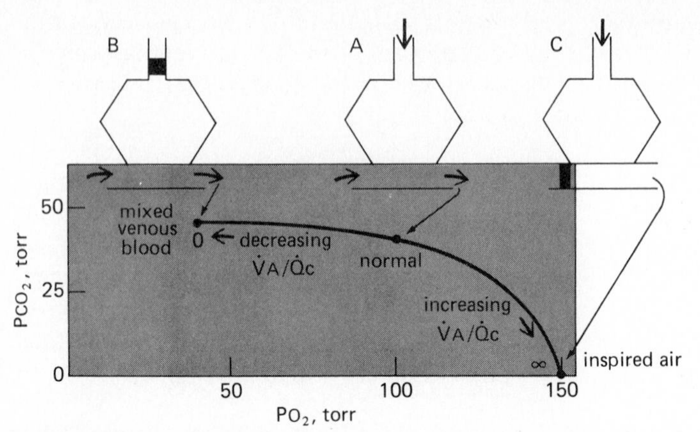

Figure 5-2 The ventilation-perfusion ratio line on an O_2-CO_2 diagram. Unit with a $\dot{V}A/\dot{Q}C$ of zero has the PO_2 and PCO_2 of mixed venous blood; a unit with an infinite $\dot{V}A/\dot{Q}C$ has the PO_2 and PCO_2 of inspired air. (*Reproduced with permission from West, 1977a.*)

tion of the ventilation-perfusion ratio line is altered if the partial pressures of the inspired gas or mixed venous blood are altered.

TESTING FOR NONUNIFORM DISTRIBUTION OF INSPIRED GAS AND PULMONARY BLOOD FLOW

Nonuniform ventilation of the alveoli can be caused by uneven *resistance* to airflow or nonuniform *compliance* in different parts of the lung. Uneven resistance to airflow may be a result of collapse of airways, as seen in emphysema; bronchoconstriction, as in asthma; decreased lumen diameter due to inflammation, as in bronchitis; obstruction by mucus, as in asthma or chronic bronchitis; or compression by tumors or edema. Uneven compliance may be a result of fibrosis, regional variations in surfactant production, pulmonary vascular congestion or edema, emphysema, diffuse or regional atelectasis, pneumothorax, or compression by tumors or cysts.

Nonuniform perfusion of the lung can be caused by embolization or thrombosis; compression of pulmonary vessels by high alveolar pressures, tumors, exudates, edema, pneumothorax, or hydrothorax; destruction or occlusion of pulmonary vessels by various disease processes; pulmonary vascular hypotension; or collapse or overexpansion of alveoli.

As already noted in Chaps. 3 and 4, gravity and regional differences in intrapleural pressure cause a degree of nonuniformity in the distribution of ventilation and perfusion in normal lungs. This will be discussed in detail later in this chapter.

Testing for Nonuniform Distribution of Inspired Gas

Several methods can be used to demonstrate an abnormal distribution of ventilation in a patient.

Single-Breath-of-Oxygen Test A rising expired nitrogen concentration in phase III (the "alveolar plateau") of the single-breath-of-oxygen test shown in Fig. 3-13 indicates the possibility of a maldistribution of ventilation, as explained in Chap. 3.

Nitrogen-Washout Test The same equipment used in the single-breath-of-oxygen test mentioned above can be used in another test for nonuniform ventilation of the lungs. In this test the subject breathes normally from a bag of 100% oxygen and his or her expired nitrogen concentration is monitored over a number of breaths. With each successive inspiration of 100% oxygen and subsequent expiration, the expired end-tidal nitrogen concentration falls as nitrogen is washed out of the lung, as shown schematically in Fig. 5-3.

The rate of decrease of the expired end-tidal nitrogen concentration depends on several factors. A high FRC, a low tidal volume, a large dead space, and a low breathing frequency will each contribute to a slower washout of alveolar nitrogen. Nonetheless, subjects with a normal distribution of airways resistance will reduce their expired end-tidal nitrogen concentration to less than 2.5% within 7 min. Subjects breathing normally who take more than 7 min. to reach an alveolar nitrogen concentration of less than 2.5% have high-resistance pathways or "slow alveoli," as discussed near the end of Chap. 2.

If the logarithms of the end-tidal nitrogen concentrations are plotted against the number of breaths taken (for a subject breathing regularly), a straight line results, as shown in Fig. 5-4*A*. (Many nitrogen meters have a log $[N_2]$ output.) On the other hand, the log $[N_2]$ plotted for a patient with a maldistribution of airways resistance, such as that produced experimen-

Figure 5-3 Schematic representation of a nitrogen-washout curve.

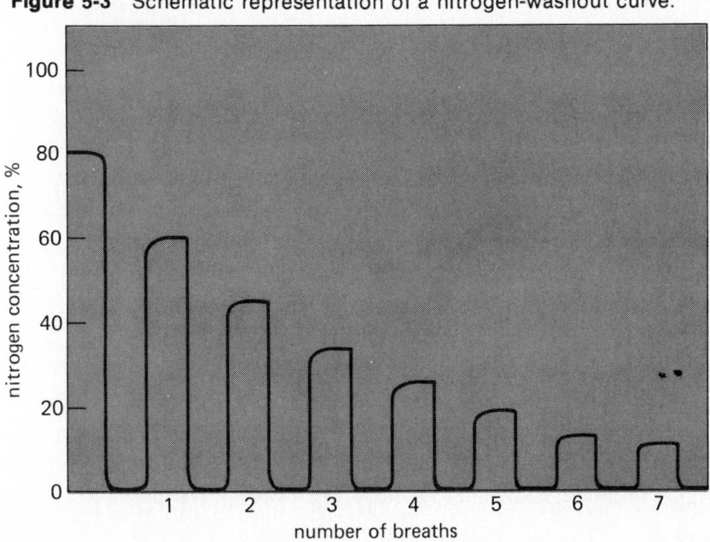

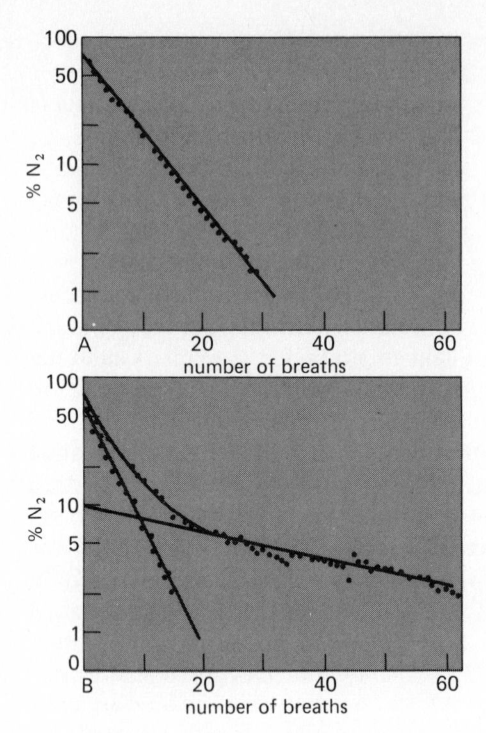

Figure 5-4 Expired nitrogen concentration versus number of breaths during a nitrogen washout. Note the logarithmic scale for the nitrogen concentration. *A*. Curve from a normal subject. *B*. Curve from a normal subject after inhalation of a histamine aerosol, which produces a marked nonuniformity of ventilation. (*Reproduced with permission from Bouhuys, 1960.*)

tally by inhaling a histamine aerosol, as shown in Fig. 5-4*B*, displays a more complex curve. After a short period of relatively rapid nitrogen washout, a long period of extremely slow nitrogen washout occurs, indicating a population of poorly ventilated "slow alveoli."

Trapped Gas Differences between the FRC determined by the helium dilution technique and the FRC determined using a body plethysmograph may indicate gas trapped in the alveoli because of airway closure, as discussed in Chap. 3. In fact, if a number of high-resistance pathways are present in the lung of the patient being tested, it may take an exceptionally long time for the patient's expired end-tidal helium concentration to equilibrate with the helium concentration in the spirometer. The closing volume determination, discussed at the end of Chap. 3, can also demonstrate airway closure in the lung.

Radioactive Xenon The methods described thus far can indicate the presence of poorly ventilated regions of the lung but not their location. Pictures of the whole lung taken with a scintillation counter, after the subject has taken a breath of a radioactive gas mixture such as ^{133}Xe and O_2, can indicate which regions of the lung are poorly ventilated.

Testing for Nonuniform Distribution of Pulmonary Blood Flow

These methods were all discussed briefly in Chap. 4. They include *angiograms,* lung scans after intravenous injection of radiolabeled *macroaggregates of albumin,* and lung scans after intravenous administration of ^{133}Xe. Each of these methods can indicate the locations of relatively large regions of poor perfusion.

Testing for Mismatched Ventilation and Perfusion

Several methods can demonstrate the presence or location of areas of the lung with mismatched ventilation and perfusion. These methods include calculations of the physiological shunt, the physiological dead space, differences between the alveolar and arterial P_{O_2}'s and P_{CO_2}'s, and lung scans after inhaled and intravenously administered ^{133}Xe.

Physiological Shunts and the Shunt Equation A right-to-left shunt is the mixing of venous blood that has not been oxygenated (or not fully oxygenated) into the arterial blood. The *physiological shunt,* which corresponds to the physiological dead space, consists of the anatomic shunts plus the intrapulmonary shunts. The intrapulmonary shunts can be *absolute shunts* (like the anatomic shunts) or they can be *"shuntlike states,"* i.e., areas of low ventilation-perfusion ratios in which alveoli are overperfused.

Anatomic Shunts Anatomic shunts consist of systemic venous blood entering the left ventricle without having entered the pulmonary vasculature. In a normal healthy adult about 2 to 5 percent of the cardiac output, including venous blood from the bronchial veins, the thebesian veins, the anterior cardiac veins, and the pleural veins, enters the left side of the circulation directly without passing through the pulmonary capillaries. (Unfortunately, this normal anatomic shunt is also occasionally referred to as the *physiological shunt* because it does not represent a pathological condition.) Pathological anatomic shunts such as right-to-left intracardiac shunts can also occur, as in tetralogy of Fallot.

Absolute Intrapulmonary Shunts Mixed venous blood perfusing pulmonary capillaries associated with totally unventilated or collapsed alveoli constitutes an absolute shunt (like the anatomic shunts) because no gas exchange occurs as the blood passes through the lung. Absolute shunts are sometimes also referred to as *true shunts.*

Shuntlike States Alveolar-capillary units with low $\dot{V}_A/\dot{Q}_C$'s also act to lower the arterial oxygen content because blood draining these units has a lower P_{O_2} than blood from units with well-matched ventilation and perfusion.

The Shunt Equation The shunt equation *conceptually* divides all alveolar-capillary units into two groups: those with well-matched ventilation and perfusion and those units with ventilation-perfusion ratios of zero. Thus, the shunt equation combines the areas of absolute shunt (including the anatomic shunts) and the shuntlike areas into a single conceptual group. The resulting ratio of shunt flow to the cardiac output, often referred to as the *venous admixture,* is the part of the cardiac output that would have to be perfusing *absolutely unventilated alveoli* to cause the systemic arterial oxygen content obtained from a particular patient. A much larger portion of the cardiac output could be *overperfusing* poorly ventilated alveoli and yield the same ratio.

The shunt equation (sometimes called the *mixing equation*) can be derived as follows: Let $\dot{Q}t$ represent the total pulmonary blood flow per minute (i.e., the cardiac output), and let $\dot{Q}s$ represent the amount of blood flow per minute entering the systemic arterial blood without receiving *any* oxygen (the "shunt flow"). The volume of blood per minute that perfuses alveolar-capillary units with well-matched ventilation and perfusion then equals $\dot{Q}t - \dot{Q}s$.

The total volume of oxygen per time entering the systemic arteries is therefore

$$\dot{Q}t \times Ca_{O_2}$$

where Ca_{O_2} = oxygen *content* of arterial blood, ml O_2 per 100 ml blood

This total amount of oxygen per time entering the systemic arteries is composed of the oxygen coming from the well-ventilated and well-perfused alveolar-capillary units:

$$(\dot{Q}t - \dot{Q}s) \times Cc'_{O_2}$$

where Cc'_{O_2} equals the oxygen content of the blood at the end of the ventilated and perfused pulmonary capillaries, *plus* the oxygen in the unaltered mixed venous blood coming from the shunt, $\dot{Q}s \times C\bar{v}_{O_2}$ (where $C\bar{v}_{O_2}$ is equal to the oxygen content of the mixed venous blood).

That is,

$$\underset{\substack{\text{Oxygen delivery} \\ \text{to systemic arteries}}}{\dot{Q}t \times Ca_{O_2}} = \underset{\substack{\text{Oxygen coming} \\ \text{from normal} \\ \dot{V}A/\dot{Q}c \text{ units}}}{(\dot{Q}t - \dot{Q}s) \times Cc'_{O_2}} + \underset{\substack{\text{Oxygen from} \\ \text{shunted blood} \\ \text{flow}}}{\dot{Q}s \times C\bar{v}_{O_2}}$$

$$\dot{Q}t \times Ca_{O_2} = \dot{Q}t \times Cc'_{O_2} - \dot{Q}s \times Cc'_{O_2} + \dot{Q}s \times C\bar{v}_{O_2}$$

$$\dot{Q}s \times (Cc'_{O_2} - C\bar{v}_{O_2}) = \dot{Q}t \times (Cc'_{O_2} - Ca_{O_2})$$

$$\frac{\dot{Q}s}{\dot{Q}t} = \frac{Cc'_{O_2} - Ca_{O_2}}{Cc'_{O_2} - C\bar{v}_{O_2}}$$

The shunt fraction, $\dot{Q}s/\dot{Q}t$, is usually multiplied by 100% so that the shunt flow is expressed as a percent of the cardiac output.

The arterial and mixed venous oxygen contents can be determined if blood samples are obtained from a systemic artery and from the pulmonary artery (for mixed venous blood), but the oxygen content of the blood at the end of the pulmonary capillaries with well-matched ventilation and perfusion is, of course, impossible to measure directly. This must be calculated from the *alveolar air equation,* discussed in Chap. 3, and the patient's hemoglobin concentration, as will be discussed in Chap. 7.

The relative contributions of the true intrapulmonary shunts and the shuntlike states to the calculated shunt flow can be estimated by repeating the measurements and calculations with the patient on a normal or slightly elevated inspired concentration of oxygen and then on a very high inspired oxygen concentration ($F_{I_{O_2}}$'s of 0.95 to 1.00). At the lower inspired oxygen concentrations the calculated $\dot{Q}s/\dot{Q}t$ will include both the true shunts and the alveolar-capillary units with low ventilation-perfusion ratios. Nevertheless, after a patient has inspired nearly 100% oxygen for 20 to 30 min, even alveoli with very low $\dot{V}_A/\dot{Q}c$'s will have high enough alveolar P_{O_2}'s to completely saturate the hemoglobin in the blood perfusing them. These units will therefore no longer contribute to the calculated $\dot{Q}s/\dot{Q}t$ and the new calculated shunt should include only areas of absolute shunt. Very high inspired oxygen concentrations may lead to absorption atelectasis of very poorly ventilated alveoli that remain perfused, and so this test may alter what it is trying to measure if high $F_{I_{O_2}}$'s are used.

Physiological Dead Space The use of the Bohr equation to determine the physiological dead space was discussed in detail in Chap. 3. If the anatomic dead space is subtracted from the physiological dead space, the result (if there is a difference) is *alveolar dead space,* or areas of infinite ventilation-perfusion ratios. Alveolar dead space also results in an alveolar-arterial CO_2 difference.

Alveolar-Arterial Oxygen Difference Larger-than-normal differences between the alveolar and arterial P_{O_2} may indicate a degree of ventilation-perfusion mismatch; however, the alveolar-arterial oxygen difference, normally 5 to 15 torr, is not specifically caused only by ventilation-perfusion mismatch. It may also be caused by anatomic shunts or diffusion block.

Single-breath Carbon Dioxide Test The expired concentration of carbon dioxide can be monitored by a rapid-response CO_2 meter in a manner similar to that used in the single-breath tests utilizing a nitrogen meter, as described in Chap. 3. The alveolar plateau phase of the expired CO_2 concentration *may* show signs of poorly matched ventilation and perfusion if such regions empty asynchronously with other regions of the lung.

Lung Scans After Radioactive Xenon Lung scans after both inhaled and injected ^{133}Xe can be used to inspect the location and amount of ventilation and perfusion to the various regions of the lung, as described previously in Chaps. 3 and 4. Regional ventilation-perfusion relationships can then be determined.

REGIONAL $\dot{V}_A/\dot{Q}_C$ DIFFERENCES AND THEIR CONSEQUENCES IN THE LUNG

The regional variations in ventilation in the normal upright lung were discussed in Chap. 3. Gravity-dependent regions of the lung receive more ventilation per unit volume than do upper regions of the lung when one is breathing near the FRC. The reason for this regional difference in ventilation is that there is a gradient of pleural surface pressure, which is probably caused by gravity and the mechanical interaction of the lung and the chest wall. The pleural surface pressure is more negative in nondependent regions of the lung, and so the alveoli in these areas are subjected to greater transpulmonary pressures. As a result, these alveoli have larger volumes than alveoli in more dependent regions of the lung and are therefore on a less steep portion of their pressure-volume curves. These less compliant alveoli change their volume less with each breath than do those in more dependent regions.

The more gravity-dependent regions of the lung also receive more blood flow per unit volume than do the upper regions of the lung, as discussed in Chap. 4. The reason for this is that the intravascular pressure in the lower regions of the lung is greater because of hydrostatic effects. Blood vessels in more dependent regions of the lung are therefore more distended or more vessels are perfused because of recruitment.

Regional Differences in Ventilation-Perfusion Ratios in the Upright Lung

Simplified graphs of the gradients of ventilation and perfusion from the bottom to the top of normal upright dog lungs are shown plotted on the same axes in Fig. 5-5. The ventilation-perfusion ratio was then calculated for several locations.

Figure 5-5 shows that the gradient of perfusion from the bottom of the lung to the top is greater than the gradient of ventilation. Because of this, the ventilation-perfusion ratio is relatively low in more gravity-dependent regions of the lung and higher in upper regions of the lung. In fact, if pulmonary perfusion pressure is low, for example, because of hemorrhage, or if alveolar pressure is high, because of positive-pressure ventila-

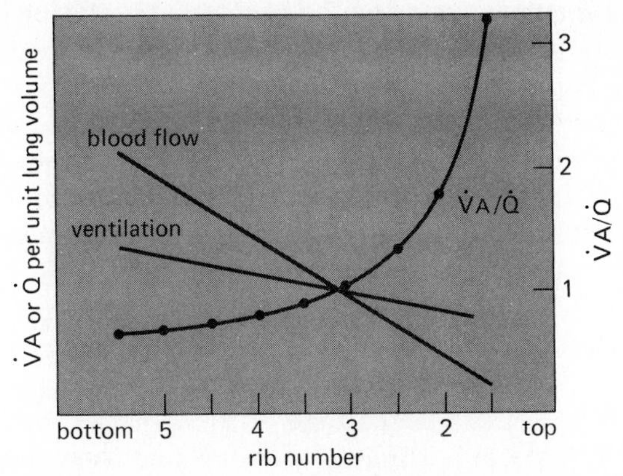

Figure 5-5 Distribution of ventilation and perfusion and ventilation-perfusion ratio down the upright dog lung. (*Reproduced with permission from West, 1977a.*)

tion with positive end-expiratory pressure, or if both factors are present, then there may be areas of zone 1 in the upper parts of the lung with infinite ventilation-perfusion ratios.

The Consequences of Regional Ventilation-Perfusion Differences in the Normal Upright Lung

The effects of the regional differences in $\dot{V}A/\dot{Q}c$ on the alveolar Po_2 and Pco_2 can be seen in Fig. 5-6. The lung was arbitrarily divided into nine imaginary horizontal sections and the $\dot{V}A/\dot{Q}c$ was calculated for each sec-

Figure 5-6 The $\dot{V}A/\dot{Q}c$ ratios for each of nine imaginary sections of a vertical lung on the ventilation-perfusion line of an O_2-CO_2 diagram. Upper sections have higher $\dot{V}A/\dot{Q}c$'s with higher Po_2's and lower Pco_2's, and lower sections have lower $\dot{V}A/\dot{Q}c$'s with lower Po_2's and higher Pco_2's. (*Reproduced with permission from West, 1977a.*)

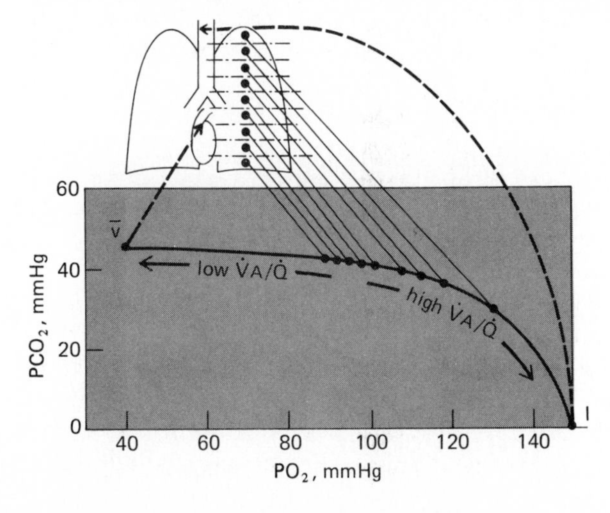

tion. These sections were then positioned on the ventilation-perfusion line of the O_2-CO_2 diagram and the P_{O_2} and P_{CO_2} of the alveoli in each section could then be estimated. Under normal circumstances the blood in the pulmonary capillaries equilibrates with the alveolar P_{O_2} and P_{CO_2} as it travels through the lung, and so the effects of regional differences in $\dot{V}_A/\dot{Q}_c$ on the regional gas exchange could be predicted. As can be seen from the figure, the upper sections have relatively high P_{O_2}'s and low P_{CO_2}'s; the lower sections have relatively low P_{O_2}'s and high P_{CO_2}'s. The results of such predictions are shown in Fig. 5-7. Only the bottom and top sections are shown, for the sake of clarity.

The lower regions of the lung receive both better ventilation and better perfusion than upper portions of the lung. The perfusion gradient is much steeper than the ventilation gradient, however, as was seen in Fig. 5-5, and so the ventilation-perfusion ratio is much higher (3.3) in the apical section than it is in the basal section (0.63). As a result, the alveolar P_{O_2} is higher (132 torr) and the alveolar P_{CO_2} is lower (28 torr) in the apical section than those in the basal section (89 and 42 torr, respectively). This means that the oxygen *content* of the blood draining the apical section is higher (20 ml O_2 per 100 ml blood) and the carbon dioxide content is lower

Figure 5-7. Regional differences in ventilation, perfusion, ventilation-perfusion ratio, gas partial pressures, and gas exchange in imaginary sections of a normal lung. Only the apical and basal sections are shown for clarity. (*Reproduced with permission from West, 1979.*)

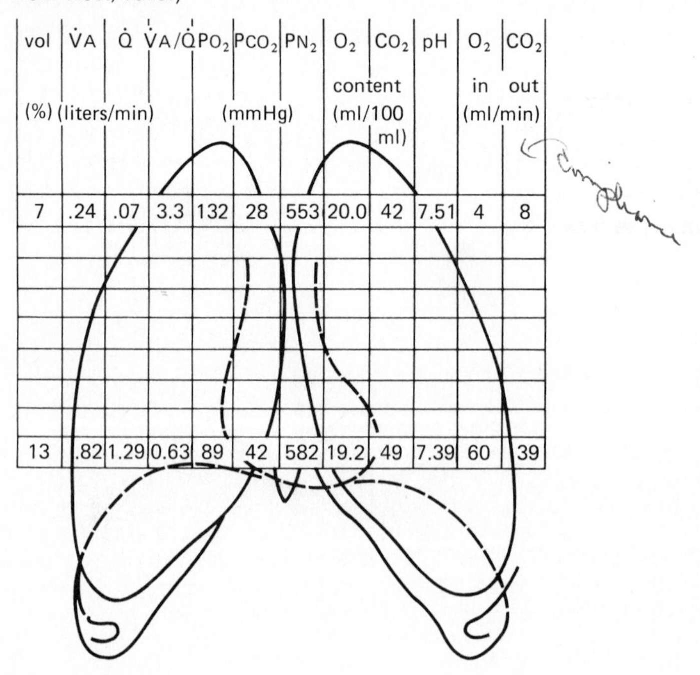

(42 ml CO_2 per 100 ml blood) than those of the blood draining the basal section (19.2 ml O_2 per 100 ml blood and 49 ml CO_2 per 100 ml blood, respectively). These contents are based, however, on *milliliters of blood* (see Chap. 7), and the blood flow to the apical section is *much lower* than the blood flow to the basal section. Therefore, even though the uppermost sections have the highest $\dot{V}_A/\dot{Q}_c$'s and P_{O_2}'s and the lowest P_{CO_2}'s, there is more *gas exchange* in the more basal sections, as the calculations in the last two columns demonstrate.

THE CAUSES OF HYPOXIA

If large amounts of the lung are in a state of ventilation-perfusion mismatch, then gas exchange between the alveoli and the pulmonary capillaries will be impaired. This impairment is likely to lead to a low arterial P_{O_2}. Because the arterial P_{O_2} is the major determinant of the combination of oxygen with hemoglobin, as will be discussed in Chap. 7, the oxygen content of the arterial blood (in milliliters of O_2 per 100 ml of blood) will also be reduced, which could lead to tissue hypoxia.

At this point the various causes of tissue hypoxia should be considered. These causes can be classified (in some cases rather arbitrarily) into four or five major groups, as shown in Table 5-1. The underlying physiology of some of these types of hypoxia has already been discussed in this or previous chapters; others will be discussed in greater detail in subsequent chapters.

Hypoxic Hypoxia

Hypoxic hypoxia refers to conditions in which the arterial P_{O_2} is abnormally low. Because the amount of oxygen that will combine with hemoglobin is mainly determined by the P_{O_2}, such conditions may lead to decreased oxygen delivery to the tissues if reflexes or other responses cannot adequately increase the cardiac output or hemoglobin concentration of the blood.

Table 5-1 A Classification of the Causes of Hypoxia

Classification	$P_{A_{O_2}}$	Pa_{O_2}	Ca_{O_2}	$P\bar{v}_{O_2}$	$C\bar{v}_{O_2}$	Increased $F_{I_{O_2}}$ helpful?
Hypoxic hypoxia						
Low alveolar P_{O_2}	Low	Low	Low	Low	Low	Yes
Diffusion impairment	Normal	Low	Low	Low	Low	Yes
Right-to-left shunts	Normal	Low	Low	Low	Low	No
$\dot{V}/\dot{Q}$ mismatch	Normal	Low	Low	Low	Low	Yes
Anemic hypoxia	Normal	Normal	Low	Low	Low	No
CO poisoning	Normal	Normal	Low	Low	Low	Possibly
Hypoperfusion hypoxia	Normal	Normal	Normal	Low	Low	No
Histotoxic hypoxia	Normal	Normal	Normal	High	High	No

Low Alveolar Po_2 Conditions causing low alveolar Po_2's inevitably lead to low arterial Po_2's and oxygen contents because the alveolar Po_2 determines the upper limit of arterial Po_2. *Hypoventilation* leads to both alveolar hypoxia and hypercapnia (high CO_2), as discussed in Chap. 3. Hypoventilation can be caused by depression or injury of the respiratory centers in the brain (discussed in Chap. 9); interference with the nerves supplying the respiratory muscles, as in myasthenia gravis; and altered mechanics of the lung or chest wall, as in noncompliant lungs because of sarcoidosis, reduced chest wall mobility because of kyphoscoliosis or obesity, and in airway obstruction. Ascent to *high altitudes* causes alveolar hypoxia because of the reduced total barometric pressure encountered above sea level. *Reduced $F_{I_{O_2}}$'s* have similar effects. Alveolar CO_2 is decreased because of the reflex increase in ventilation caused by hypoxic stimulation, as will be discussed in Chap. 11. Hypoventilation and ascent to high altitudes lead to decreased venous Po_2 and oxygen content as oxygen is extracted from the already hypoxic arterial blood. Administration of elevated oxygen concentrations in the inspired gas ($F_{I_{O_2}}$'s) can alleviate the alveolar and arterial hypoxia in hypoventilation and in ascent to high altitude, but it cannot reverse the hypercapnia of hypoventilation. In fact, administration of elevated $F_{I_{O_2}}$'s to spontaneously breathing patients hypoventilating because of a depressed central response to CO_2 (see Chap. 9) can further depress ventilation.

Diffusion Impairment Alveolar-capillary diffusion will be discussed in greater detail in the next chapter. Conditions such as interstitial fibrosis and interstitial or alveolar edema can lead to low arterial Po_2's and contents with normal or elevated alveolar Po_2's. High $F_{I_{O_2}}$'s that increase the alveolar Po_2 to very high levels may raise the arterial Po_2 by increasing the partial pressure gradient for oxygen diffusion, as will be seen in Chap. 6.

Shunts True right-to-left shunts, such as anatomic shunts and absolute intrapulmonary shunts, can cause decreased arterial Po_2's with normal or even elevated alveolar Po_2's. Arterial hypoxia caused by true shunts is not relieved by high $F_{I_{O_2}}$'s because the shunted blood does not come into contact with the high levels of oxygen. The hemoglobin of the *unshunted* blood is nearly completely saturated with oxygen at a normal $F_{I_{O_2}}$ of 0.21, and the small additional volume of oxygen dissolved in the blood at high $F_{I_{O_2}}$'s cannot make up for the low hemoglobin saturation of the shunted blood, as will be discussed in Chap. 7.

$\dot{V}_A/\dot{Q}_c$ Mismatch Alveolar-capillary units with low ventilation-perfusion ratios contribute to arterial hypoxia, as already discussed. Units with high $\dot{V}_A/\dot{Q}_c$'s do not by themselves lead to arterial hypoxia, of course, but large lung areas that are underperfused are usually associated with either overperfusion of other units or with low cardiac outputs (see "Hypoperfusion Hypoxia" below). Hypoxic pulmonary vasoconstriction (discussed in Chap. 4) and local airway reflexes (discussed in Chap. 2) normally help to minimize ventilation-perfusion mismatch.

Anemic Hypoxia

Anemic hypoxia is caused by a decrease in the amount of functioning hemoglobin, which can be a result of decreased hemoglobin or erythrocyte production, the production of abnormal hemoglobin or red blood cells, pathological destruction of erythrocytes, or interference with the chemical combination of oxygen and hemoglobin. Carbon monoxide poisoning, for example, results from the greater affinity of hemoglobin for carbon monoxide than for oxygen. Methemoglobinemia is a condition in which the iron in hemoglobin has been altered from the Fe^{2+} to the Fe^{3+} form, which does not combine with oxygen. The transport of oxygen by hemoglobin will be discussed in greater detail in Chap. 7.

Anemic hypoxia results in a decreased oxygen content when both alveolar and arterial P_{O_2} are normal. Venous P_{O_2} and oxygen content are both decreased. Administration of high $F_{I_{O_2}}$'s is not effective in greatly increasing the arterial oxygen content (except possibly in carbon monoxide poisoning).

Hypoperfusion Hypoxia

Hypoperfusion hypoxia (sometimes called *stagnant hypoxia*) results from low blood flow. This can occur either locally, in a particular vascular bed, or systemically, in the case of a low cardiac output. The alveolar P_{O_2} and the arterial P_{O_2} and oxygen content may be normal, but the reduced oxygen delivery to the tissues may result in tissue hypoxia. Venous P_{O_2} and oxygen content are low. Raising the $F_{I_{O_2}}$ is of little value in hypoperfusion hypoxia (unless it directly increases the perfusion) because the blood flowing to the tissues is already oxygenated normally.

Histotoxic Hypoxia

Histotoxic hypoxia refers to a poisoning of the cellular machinery that uses oxygen to produce energy. Cyanide, for example, binds to cytochrome oxidase in the respiratory chain and effectively blocks oxidative phosphorylation. Alveolar P_{O_2} and arterial P_{O_2} and oxygen content may be normal (or even *elevated,* because low doses of cyanide increase ventilation by stimulating the arterial chemoreceptors). Venous P_{O_2} and oxygen content are elevated because oxygen is not utilized.

Other Causes of Hypoxia

Tissue edema or fibrosis may result in impaired diffusion of oxygen from the blood to the tissues. It is also conceivable that the delivery of oxygen to a tissue is completely normal, but the tissue's metabolic demands still exceed the supply and tissue hypoxia could result. This is known as *overutilization hypoxia*.

The Effects of Hypoxia

Hypoxia can result in reversible tissue injury or even tissue death. The outcome of an hypoxic episode depends on whether the tissue hypoxia is generalized or localized, on how severe the hypoxia is, on the rate of development of the hypoxia (see Chap. 11), and on the duration of the hypoxia. Different cell types have different susceptibilities to hypoxia; unfortunately, brain cells and heart cells are the most susceptible.

Chapter 6

Diffusion of Gases

OBJECTIVES

The student understands the diffusion of gases in the lung.

1 Defines diffusion and distinguishes it from "bulk flow."
2 States Fick's law for diffusion.
3 Distinguishes between perfusion limitation and diffusion limitation of gas transfer in the lung.
4 Describes the diffusion of oxygen from the alveoli into the blood.
5 Describes the diffusion of carbon dioxide from the blood to the alveoli.
6 Defines the diffusing capacity and discusses its measurement.

Diffusion of a gas occurs when there is a net movement of molecules from an area in which *that particular gas* exerts a high partial pressure to an area in which it exerts a lower partial pressure. Movement of a gas by diffusion is therefore different from the movement of gases through the conducting airways, which occurs by "bulk flow" (mass movement or convection). During bulk flow gas movement is due to differences in *total* pressure, and molecules of different gases move together along the total pressure gradient. During diffusion, different gases each move according to their own individual partial pressure gradients. Gas transfer during diffusion occurs by random molecular movement. It is therefore dependent on temperature because molecular movement increases at higher temperatures. Gas movement occurs in both directions during diffusion, but the area of higher partial pressure, because of its greater number of molecules per unit volume, has proportionately more random "departures." Thus, the *net* movement of gas is dependent on the partial pressure difference between the two areas. In a static situation diffusion continues until no partial pressure differences exist for any gases in the two areas; in the lungs oxygen and carbon dioxide continually enter and leave the alveoli, and so such an equilibrium does not take place.

FICK'S LAW FOR DIFFUSION

The diffusion of gases in the lung is a multistep operation. For example, oxygen is brought into the alveoli by bulk flow through the conducting airways. When air flows through the conducting airways during inspiration, the linear velocity of the bulk flow decreases as the air approaches the alveoli. This is because the total cross-sectional area increases dramatically in the distal portions of the tracheobronchial tree, as was seen in Fig. 1-5. By the time the air reaches the alveoli, bulk flow probably ceases and further gas movement occurs by diffusion. Oxygen then moves through the gas phase in the alveoli according to its own partial pressure gradient. The distance from alveolar duct to the alveolar-capillary interface is probably less than 1 mm. Diffusion in the alveolar gas phase is believed to be greatly assisted by the pulsations of the heart and blood flow, which are transmitted to the alveoli and increase molecular motion.

Oxygen then diffuses through the alveolar-capillary interface. It must first, therefore, move from the gas phase to the liquid phase, according to Henry's law, which states that the amount of a gas absorbed by a liquid with which it does not combine chemically is directly proportional to the partial pressure of the gas to which the liquid is exposed and the solubility of the gas in the liquid. Oxygen must dissolve in and diffuse through the thin layer of pulmonary surfactant, the alveolar epithelium, the interstitium, and the capillary endothelium, as was shown in Fig. 1-4 (step 2, near the arrow). It must then diffuse through the plasma (step 3), where some remains dissolved and the majority enters the erythrocyte and combines with hemoglobin (step 4). The blood then carries the oxygen out of the lung by bulk flow and distributes it to the other tissues of the body, as was shown in Fig. 1-1. At the tissues, oxygen diffuses from the erythrocyte through the plasma, capillary endothelium, interstitium, tissue cell membrane, cell interior, and into the mitochondrial membrane. The process is almost entirely reversed for carbon dioxide, as shown in Fig. 1-1.

The factors that determine the rate of diffusion of a gas through the alveolar-capillary barrier are described by *Fick's law for diffusion,* shown here in a simplified form:

$$\dot{V}_{gas} = \frac{A \times D \times (P_1 - P_2)}{T}$$

where $\dot{V}_{gas}$ = volume of gas diffusing through the tissue barrier per time, ml/min
A = surface area of the barrier available for diffusion
D = diffusion coefficient, or diffusivity, of the particular gas in the barrier
T = thickness of the barrier or the diffusion distance
$P_1 - P_2$ = partial pressure difference of the gas across the barrier

That is, the volume of gas per unit of time moving across the alveolar-capillary barrier is directly proportional to the area of the barrier, the diffusivity, and the difference in concentration between the two sides, but is inversely proportional to the barrier thickness.

The surface area of the blood-gas barrier is believed to be at least 70 m^2 in a healthy average-sized adult at rest. That is, about 70 m^2 of the *potential* surface area is both ventilated and perfused at rest. If more capillaries are recruited, as in exercise, the surface area available for diffusion increases; if venous return falls, for example, because of hemorrhage, or if alveolar pressure is raised by positive-pressure ventilation, then capillaries may be derecruited and the surface area available for diffusion may decrease.

The thickness of the alveolar-capillary diffusion barrier is only about 0.2 to 0.5 μm. This barrier thickness can increase in interstitial fibrosis or interstitial edema, thus interfering with diffusion. Diffusion probably increases at higher lung volumes because as alveoli are stretched the diffusion distance decreases slightly (and also because small airways subject to closure may be open at higher lung volumes).

The diffusivity, or diffusion constant, for a gas is directly proportional to the solubility of the gas in the diffusion barrier and inversely proportional to the square root of the density of the gas:

$$D \propto \frac{\text{solubility}}{\sqrt{\text{MW}}}$$

The relationship between solubility and diffusion through the barrier has already been discussed. The diffusivity is inversely proportional to the square root of the molecular weight of the gas because different gases with equal numbers of molecules in equal volumes have the same molecular energy if they are at the same temperature. Therefore, light molecules travel faster, have more frequent collisions, and diffuse more rapidly. Thus, Graham's law states that for two gases, if all else is equal, their relative rates of diffusion are inversely proportional to the square roots of their densities.

For the two gases of greatest interest in the lung,

$$\frac{\sqrt{\text{MW of } O_2}}{\sqrt{\text{MW of } CO_2}} = 0.85$$

Since the relative diffusion rates are *inversely* proportional to the ratio of the square roots of their molecular weights,

$$\frac{\text{Diffusion rate for } O_2}{\text{Diffusion rate for } CO_2} \propto \frac{1}{0.85} = 1.17$$

That is, because oxygen is less dense than carbon dioxide, it should diffuse 1.2 times as fast as carbon dioxide (which it does as it moves through the alveoli). In the alveolar-capillary barrier, however, the relative solubilities of oxygen and carbon dioxide must also be considered. The solubility of carbon dioxide in the liquid phase is about 24 times that of oxygen, and so carbon dioxide diffuses about 0.85×24, or about *20 times,* more rapidly through the alveolar-capillary barrier than does oxygen. For this reason patients develop problems in oxygen diffusion through the alveolar-capillary barrier before carbon dioxide retention due to diffusion impairment occurs.

LIMITATIONS OF GAS TRANSFER

The factors that limit the movement of a gas through the alveolar-capillary barrier, as described by Fick's law for diffusion, can be arbitrarily divided into three components: the diffusion coefficient, the surface area and thickness of the alveolar-capillary membrane, and the partial pressure gradient across the barrier for each particular gas. The diffusion coefficient, as discussed in the previous section, is dependent on the physical properties of the gases and the alveolar-capillary membrane. The surface area and thickness of the membrane are physical properties of the barrier, but they can be altered by changes in the pulmonary capillary blood volume, the cardiac output or pulmonary artery pressure, or by changes in lung volume. The partial pressure gradient of a gas (across the barrier) is the final major determinant of its rate of diffusion. The partial pressure of a gas in the *mixed venous blood* and in the *pulmonary capillaries* is just as important a factor as its *alveolar* partial pressure in determining its rate of diffusion. This fact will be demonstrated in the next section.

Diffusion Limitation

An erythrocyte and its attendant plasma spend an average of about 0.75 to 1.2 s inside the pulmonary capillaries at resting cardiac outputs. This time can be estimated by dividing the pulmonary capillary blood volume by the pulmonary blood flow (expressed in milliliters per second). Some erythrocytes may take less time to traverse the pulmonary capillaries; others may take longer. Figure 6-1 shows schematically the calculated change with time in the partial pressures in the blood of three gases: oxygen, carbon monoxide, and nitrous oxide. These are shown in comparison to the alveolar partial pressures for each gas, as indicated by the dotted line. This alveolar partial pressure is different for each of the three gases, and it depends on its concentration in the inspired gas mixture and

Figure 6-1 Calculated changes in the partial pressure of carbon monoxide, nitrous oxide, and oxygen in the blood as it passes through a functional pulmonary capillary. There are no units on the ordinate because the scale is different for each of the three gases, depending on the alveolar partial pressure of each gas. The abscissa is in seconds, indicating the time the blood has spent in the capillary. At resting cardiac outputs blood spends an average of 0.75 s in a pulmonary capillary. The alveolar partial pressure of each gas is indicated by the dotted line. Note that the partial pressure of nitrous oxide and oxygen equilibrate rapidly with their alveolar partial pressure. (*Reproduced with permission from Comroe, 1962.*)

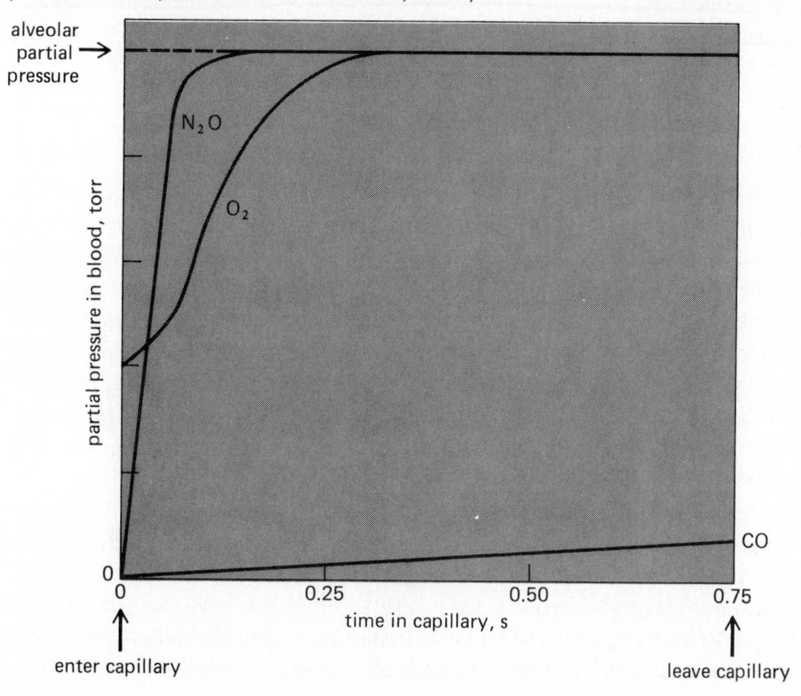

how rapidly it is removed by the pulmonary capillary blood. The schematic is drawn as though all three gases were administered simultaneously, but this is not necessarily the case. Consider each gas as though it acts independently of the others.

The partial pressure of carbon monoxide in the pulmonary capillary blood rises very slowly compared to that of the other two gases in the figure. (Obviously, a low inspired concentration of carbon monoxide must be used for a very short time in such an experiment.) Nevertheless, if the *content* of carbon monoxide (in milliliters of CO per milliliter of blood) were measured simultaneously, it would be rising very rapidly. The reason for this rapid rise is that carbon monoxide combines *chemically* with the hemoglobin in the erythrocytes. Indeed, the affinity of carbon monoxide for hemoglobin is about 210 times that of oxygen for hemoglobin. The carbon monoxide that is chemically combined with hemoglobin does not contribute to the partial pressure of carbon monoxide in the blood because it is no longer *physically dissolved* in it. Therefore, the partial pressure of carbon monoxide in the pulmonary capillary blood does not come close to the partial pressure of carbon monoxide in the alveoli during the time that the blood is exposed to the alveolar carbon monoxide. The partial pressure gradient across the alveolar-capillary barrier for carbon monoxide is thus well-maintained for the entire time the blood spends in the pulmonary capillary, and the diffusion of carbon monoxide is limited only by its diffusivity in the barrier and the surface area and thickness of the barrier. Carbon monoxide transfer from the alveolus to the pulmonary capillary blood is referred to as *diffusion-limited* rather than *perfusion-limited*.

Perfusion Limitation

The partial pressure of nitrous oxide in the pulmonary capillary blood equilibrates very rapidly with the partial pressure of nitrous oxide in the alveolus because nitrous oxide moves through the alveolar-capillary barrier very easily and because it does not combine chemically with the hemoglobin in the erythrocytes. After only about 0.1 s of exposure of the pulmonary capillary blood to the alveolar nitrous oxide, the partial pressure gradient across the alveolar-capillary barrier has been abolished. From this point on, no further nitrous oxide transfer occurs from the alveolus to that portion of the blood in the capillary that has already equilibrated with the alveolar nitrous oxide partial pressure; during the last 0.6 to 0.7 s no net diffusion occurs between the alveolus and the blood as it travels through the pulmonary capillary. Of course, blood just entering the capillary at the arterial end will not be equilibrated with the alveolar partial pressure of nitrous oxide, and so nitrous oxide can diffuse into the blood at the arterial end. The transfer of nitrous oxide is therefore *perfusion-limited*. Nitrous oxide transfer from a particular alveolus to one of its pulmonary capillaries can be increased by increasing the cardiac output and thus decreasing the amount of time the blood stays in the pulmonary capillary after equilibration of the alveolar partial pressure of nitrous oxide has occurred. (Because increasing the cardiac output may recruit previously unperfused capillaries, the *total* diffusion of both carbon monoxide and nitrous oxide may increase as the surface area for diffusion increases.)

DIFFUSION OF OXYGEN

As can be seen in Fig. 6-1, the time course for oxygen transfer falls between those for carbon monoxide and nitrous oxide. The partial pressure of oxygen rises fairly rapidly (note that it starts at the P_{O_2} of the mixed venous blood, about 40 torr, rather than at zero), and equilibration with the alveolar P_{O_2} of about 100 torr occurs within about 0.25 s, or about one-third of the time the blood is in the pulmonary capillary at normal resting cardiac outputs. Oxygen moves easily through the alveolar-capillary barrier and into the erythrocytes, where it combines chemically with hemoglobin. Oxygen does not combine with hemoglobin as quickly as does carbon monoxide, and so the partial pressure of oxygen rises more rapidly than does the partial pressure of carbon monoxide. Nonetheless, the oxygen chemically bound to hemoglobin (and therefore no longer physically dissolved) exerts no partial pressure, and so the partial pressure gradient across the alveolar-capillary membrane is initially well-maintained and oxygen transfer occurs. The chemical combination of oxygen and hemoglobin, however, occurs rapidly (within hundredths of a second), and at the normal alveolar partial pressure of oxygen, the hemoglobin becomes nearly saturated with oxygen very quickly, as will be discussed in the next chapter. As this happens, the partial pressure of oxygen in the blood rises rapidly to that in the alveolus, and from that point, no further oxygen transfer from the alveolus to the equilibrated blood can occur. Therefore, under the conditions of normal alveolar P_{O_2} and a normal resting cardiac output, oxygen transfer from alveolus to pulmonary capillary is *perfusion-limited.*

The upper portion of Fig. 6-2 shows similar graphs of calculated changes in the partial pressure of oxygen in the blood as it moves through a pulmonary capillary. The alveolar P_{O_2} is normal. During exercise, blood moves through the pulmonary capillary much more rapidly than it does at resting cardiac outputs. In fact, during severe exercise the blood may stay in the "functional" pulmonary capillary an average of only about 0.25 s, as indicated on the graph. Oxygen transfer into the blood per time will be greatly increased because there is little or no perfusion limitation of oxygen transfer. (Indeed that part of the blood that stays in the capillary *less* than the average may be subjected to diffusion limitation of oxygen trans-

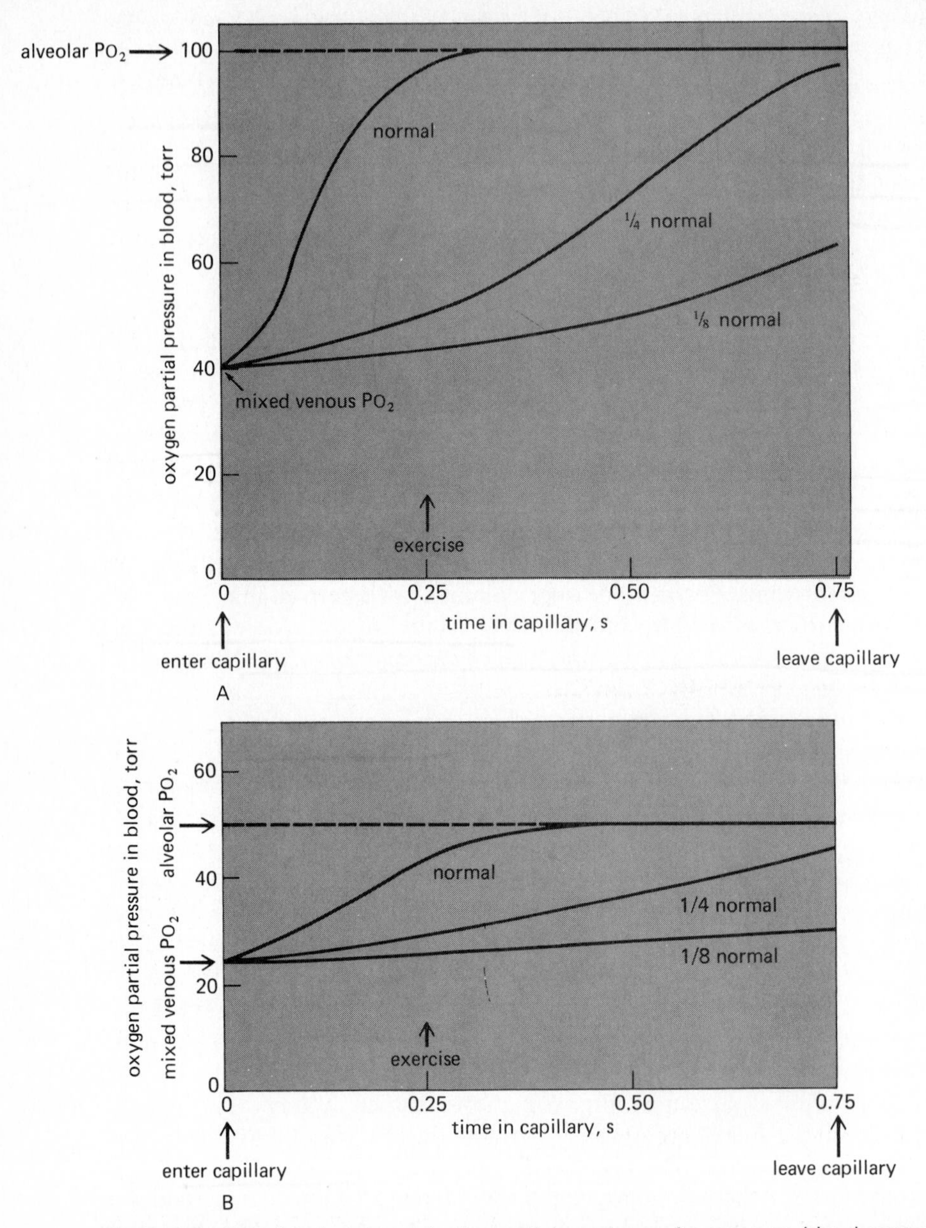

Figure 6-2 Calculated changes in the partial pressure of oxygen as blood passes through a pulmonary capillary. *Upper panel* shows the patterns with a normal alveolar P_{O_2} of about 100 torr and normal and abnormal diffusion through the alveolar-capillary barrier. *Lower panel* shows the patterns with low alveolar P_{O_2} and normal and abnormal diffusion through the alveolar-capillary barrier. Alveolar P_{O_2} is indicated by the dotted line. (*Reproduced with permission from Wagner, 1972.*)

fer.) Of course, *total* oxygen transfer is also increased during exercise because of recruitment of previously unperfused capillaries, which increases the surface area for diffusion, and because of better matching of ventilation and perfusion. A person with an abnormal alveolar-capillary barrier due to a fibrotic thickening or interstitial edema may approach diffusion limitation of oxygen transfer at rest and may have a serious diffusion limitation of oxygen transfer during strenuous exercise, as can be seen in the middle curve in Fig. 6-2*A*. A person with an extremely abnormal alveolar-capillary barrier might have diffusion limitation of oxygen transfer even at rest, as seen at right in the figure.

The effect of a low alveolar partial pressure of oxygen on oxygen transfer from the alveolus to the capillary is seen in the lower portion of Fig. 6-2. The low alveolar P_{O_2} sets the upper limit for the end-capillary blood P_{O_2}. Because the oxygen content of the arterial blood is decreased by this hypoxic hypoxia, the mixed venous P_{O_2} is also depressed. The even greater decrease in the alveolar partial pressure of oxygen, however, causes a decreased alveolar-capillary partial pressure gradient and the blood P_{O_2} takes longer to equilibrate with the alveolar P_{O_2}. For this reason, a normal person exerting himself or herself at high altitude might be subjected to diffusion limitation of oxygen transfer.

DIFFUSION OF CARBON DIOXIDE

The time course of carbon dioxide transfer from the pulmonary capillary blood to the alveolus is shown in Fig. 6-3. In a normal person with a mixed venous partial pressure of carbon dioxide of 45 torr and an alveolar partial pressure of carbon dioxide of 40 torr, an equilibrium is reached in about 0.25 s, or about the same time as that for oxygen. This fact may seem surprising, considering that the diffusivity of carbon dioxide is about 20 times that of the diffusivity of oxygen, but the partial pressure gradient is normally only about 5 torr for carbon dioxide, whereas it is about 60 torr for oxygen. Carbon dioxide transfer is therefore normally *perfusion-limited,* although it may be diffusion-limited in a person with an abnormal alveolar-capillary barrier, as shown in the figure.

MEASUREMENT OF DIFFUSING CAPACITY

It is often useful to determine the diffusion characteristics of a patient's lungs during their assessment in the pulmonary function laboratory. It may be particularly important to determine whether an apparent impairment in diffusion is a result of *perfusion limitation* or *diffusion limitation.*

The *diffusing capacity* (or *transfer factor*) is the rate at which oxygen or carbon monoxide is absorbed from the alveolar gas into the pulmonary capillaries (in milliliters per minute) per unit of partial pressure gradient

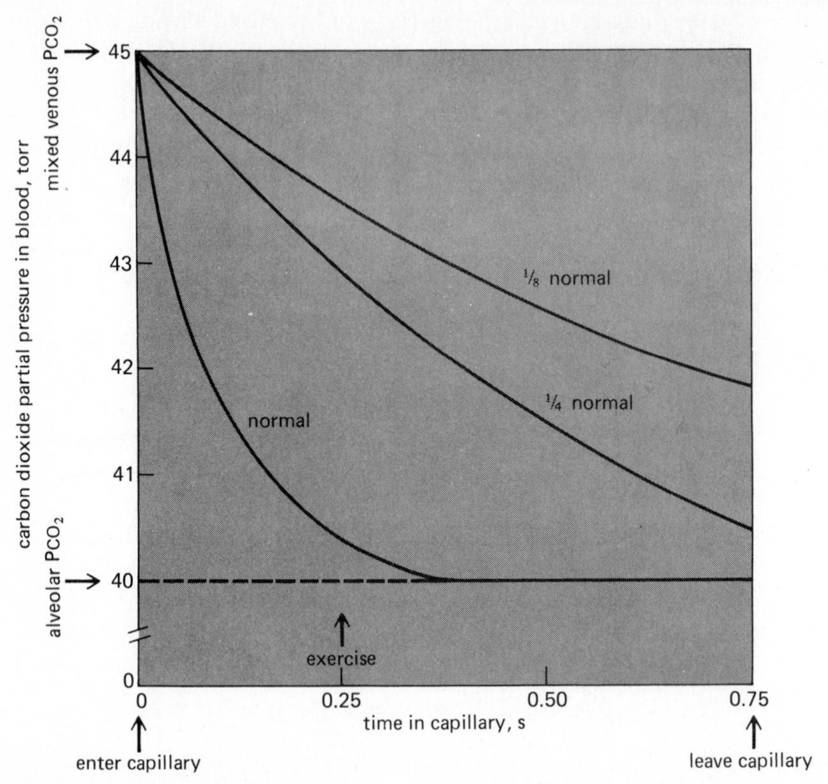

Figure 6-3 Calculated changes in the partial pressure of carbon dioxide as blood passes through a pulmonary capillary. The mixed venous P_{CO_2} is about 45 torr. The alveolar P_{CO_2} is indicated by the dotted line. Patterns for normal and abnormal diffusion through the alveolar-capillary barrier are shown. Note that the partial pressure of CO_2 in the pulmonary capillary blood normally equilibrates rapidly with the alveolar P_{CO_2}. (*Reproduced with permission from Wagner, 1972.*)

(in millimeters of mercury). The diffusing capacity of the lung (for gas x), D_{L_x}, is therefore equal to the uptake of gas x, $\dot{V}_x$, divided by the difference between the alveolar partial pressure of gas x, P_{A_x}, and the mean capillary partial pressure of gas x, $P_{\bar{c}_x}$.

$$D_{L_x} = \frac{\dot{V}_x}{P_{A_x} - P_{\bar{c}_x}} \qquad \text{ml/min/mmHg}$$

This is really just a rearrangement of the Fick equation given at the beginning of this chapter. The terms for area, diffusivity, and thickness have been combined into D_{L_x} and the equation has been rearranged:

$$\dot{V}_x = \frac{A \times D \times (P_1 - P_2)}{T}$$

$$\frac{\dot{V}_x}{P_1 - P_2} = \frac{A \times D}{T}$$

$$DL_x = \frac{\dot{V}_x}{P_1 - P_2}$$

The mean partial pressure of oxygen or carbon monoxide is, as already discussed, affected by their chemical reactions with hemoglobin, as well as by their transfer through the alveolar-capillary barrier. For this reason, the diffusing capacity of the lung is determined by both the diffusing capacity of the membrane, D_M, and the reaction with hemoglobin, expressed as $\Theta \times Vc$, where Θ is the volume of gas in milliliters per minute taken up by the erythrocytes in 1 ml of blood per millimeter of mercury partial pressure gradient between the plasma and the erythrocyte and Vc is the capillary blood volume in milliliters. (The units of $\Theta \times Vc$ are therefore ml/min/mmHg.) The diffusing capacity of the lung, D_L, can be shown to be related to D_M and $\Theta \times Vc$ as follows:

$$\frac{1}{D_L} = \frac{1}{D_M} + \frac{1}{\Theta \times Vc} \left(+ \frac{1}{D_A} \right)$$

D_A, or diffusion through the alveolus, is very rapid and may be disregarded.

Carbon monoxide is most frequently used in determinations of the diffusing capacity because the mean pulmonary capillary partial pressure of carbon monoxide is virtually zero when nonlethal alveolar partial pressures of carbon monoxide are used:

$$DL_{co} = \frac{\dot{V}CO}{PA_{co} - P\bar{c}_{co}}$$

but $P\bar{c}_{co} \simeq 0$

and so $DL_{co} = \dfrac{\dot{V}CO}{PA_{co}}$

Several different methods are used clinically to measure the carbon monoxide diffusing capacity and involve both single-breath and steady-state techniques, sometimes during exercise. The DL_{co} is decreased in diseases associated with interstitial or alveolar fibrosis, such as sarcoidosis, scleroderma, and asbestosis, or conditions causing interstitial or alveolar pulmonary edema. It is also decreased in conditions causing a decrease in the surface area available for diffusion, such as emphysema, tumors, or a low pulmonary capillary blood volume.

Chapter 7

The Transport of Oxygen and Carbon Dioxide in the Blood

OBJECTIVES

The student understands how oxygen and carbon dioxide are transported to and from the tissues in the blood.

1. States the relationship between the partial pressure of oxygen in the blood and the amount of oxygen physically dissolved in the blood.
2. Describes the chemical combination of oxygen with hemoglobin and the "oxygen dissociation curve."
3. Defines hemoglobin *saturation,* the oxygen-carrying *capacity,* and the oxygen *content* of blood.
4. States the physiological consequences of the shape of the oxygen dissociation curve.
5. Lists the physiological factors that can influence the oxygen dissociation curve and predicts their effects on oxygen transport by the blood.
6. States the relationship between the partial pressure of carbon dioxide in the blood and the amount of carbon dioxide physically dissolved in the blood.
7. Describes the transport of carbon dioxide as carbamino compounds with blood proteins.

8 Explains how most of the carbon dioxide in the blood is transported as bicarbonate.
9 Describes the carbon dioxide dissociation curve for whole blood.
10 Explains the Bohr and Haldane effects.

The final step in the exchange of gases between the external environment and the tissues is the transport of oxygen and carbon dioxide to and from the lung by the blood. Oxygen is carried both physically dissolved in the blood and chemically combined to hemoglobin. Carbon dioxide is carried physically dissolved in the blood, chemically combined to blood proteins as carbamino compounds, and as bicarbonate.

TRANSPORT OF OXYGEN BY THE BLOOD

Oxygen is transported both physically dissolved in blood and chemically combined to the hemoglobin in the erythrocytes. Much more oxygen is normally transported combined with hemoglobin than is physically dissolved in the blood. Without hemoglobin, the cardiovascular system could not supply oxygen transport adequate to meet tissue demands.

Physically Dissolved

At a temperature of 37°C, 1 ml of *plasma* contains 0.00003 ml of oxygen per torr P_{O_2}. This corresponds to Henry's law, as discussed in Chap. 6. *Whole blood* contains a similar amount of dissolved oxygen per milliliter because oxygen dissolves in the fluid of the erythrocytes in about the same amount. Therefore, normal arterial blood with a P_{O_2} of approximately 100 torr contains only about 0.003 ml of oxygen per milliliter of blood, or 0.3 ml of oxygen per 100 ml of blood. (Blood oxygen content is conventionally expressed in milliliters of O_2 per 100 ml of blood, or *volumes percent.*)

A few simple calculations can demonstrate that the oxygen physically dissolved in the blood is not sufficient to fulfill the body's oxygen demand (at normal $F_{I_{O_2}}$ and barometric pressure). The resting oxygen consumption of an adult is approximately 250 to 300 ml of oxygen per minute. If the tissues were able to remove the entire 0.3 ml of oxygen per 100 ml of blood flow they receive, the cardiac output would have to be about 83.3 liters/min to meet the tissue demand for oxygen at rest:

$$\frac{250 \text{ ml } O_2}{\text{min}} \div \frac{0.3 \text{ ml } O_2}{100 \text{ ml blood}} = \frac{83333 \text{ ml}}{\text{min}} = \frac{83.3 \text{ liters}}{\text{min}}$$

During severe exercise the oxygen demand can increase as much as 16-fold to 4 liters/min or more. Under such conditions the cardiac output would have to be greater than 1000 liters/min if physically dissolved oxygen were to supply all of the oxygen required by the tissues. The maximum cardiac outputs attainable by normal adults during severe exercise are in the range of 25 liters/min. Clearly, the *physically dissolved* oxygen in the blood cannot meet the metabolic demand for oxygen even at rest.

Chemically Combined with Hemoglobin

The Structure of Hemoglobin Hemoglobin is a complex molecule with a molecular weight of about 65,000. The protein portion (globin) has a tetrameric structure consisting of four linked polypeptide chains, each of which is attached to a protoporphyrin (heme) group. Each heme group consists of four symmetrically arranged pyrroles with a ferrous (Fe^{2+}) iron atom at its center. The iron atom is bound to each of the pyrrole groups and to one of the four polypeptide chains. A sixth binding site on the ferrous iron atom is freely available to bind with oxygen (or carbon monoxide). Therefore *each* of the four polypeptide chains can bind a molecule of oxygen (or carbon monoxide) to the iron atom in its own heme group, and so the tetrameric hemoglobin molecule can combine chemically with four oxygen molecules (or eight oxygen atoms). Both the globin component and the heme component (with its iron atom in the ferrous state), in their proper spatial orientation to each other, are necessary for the chemical reaction with oxygen to take place—neither heme nor globin alone will combine with oxygen. Each of the tetrameric hemoglobin subunits can combine with oxygen by itself, as will be discussed later in this chapter.

Variations in the amino acid sequences of the four globin subunits may have important physiological consequences. Normal adult hemoglobin (HbA) consists of two α chains, each of which has 141 amino acids, and two β chains, each of which has 146 amino acids. Fetal hemoglobin (HbF), which consists of two α chains and two γ chains, has a higher affinity for oxygen than does adult hemoglobin. Synthesis of β chains normally begins about 6 weeks before birth and HbA usually replaces almost all of the HbF by the time an infant is 4 months old. Other, *abnormal* hemoglobin molecules may be produced by genetic substitution of a single amino acid for the normal one in an α or β chain or (rarely) by alterations in the structure of heme groups. These alterations may produce changes in the affinity of the hemoglobin for oxygen, change the physical properties of hemoglobin, or alter the interaction of hemoglobin and other substances that affect its combination with oxygen, such as 2,3-diphosphoglycerate (2,3-DPG) (discussed later in this chapter). More than 120 variants of normal adult hemoglobin have been demonstrated in patients. The best known of these, hemoglobin S, is present in sickle cell anemia. Hemoglobin S is not very soluble in the cytosol of the erythrocyte when it is not combined with oxygen. It therefore may crystallize inside the red blood cell. This crystallization changes the shape of the cell from the normal biconcave disc to a crescent or "sickle" shape. A sickled cell is more fragile than a normal cell. In addition, the cells have a tendency to stick to each other, which increases blood viscosity and also favors thrombosis or blockage of blood vessels.

Chemical Reaction of Oxygen and Hemoglobin

Hemoglobin rapidly combines *reversibly* with oxygen. It is the reversibility of the reaction that allows oxygen to be released to the tissues; if the reaction did not proceed easily in both directions, hemoglobin would be of little use in delivering oxygen to satisfy metabolic needs. The reaction is very fast, with a half-time of 0.01 s or less. Each gram of hemoglobin is capable of combining with about 1.39 ml of oxygen under optimal conditions, but under normal circumstances some hemoglobin is in forms such as methemoglobin (the iron atom is in the ferric state) or is combined with carbon monoxide, in which case the hemoglobin cannot bind oxygen. For this reason, the oxygen-carrying *capacity* of hemoglobin is conventionally considered to be 1.34 ml of oxygen per gram of hemoglobin. That is, each gram of hemoglobin, when *fully saturated* with oxygen, binds 1.34 ml of oxygen. Therefore, a person with 15 g of hemoglobin per 100 ml of blood has an oxygen-carrying capacity of 20.1 ml of O_2 per 100 ml of blood:

$$\frac{15 \text{ g Hb}}{100 \text{ ml blood}} \times \frac{1.34 \text{ ml } O_2}{\text{gram Hb}} = \frac{20.1 \text{ ml } O_2}{100 \text{ ml blood}}$$

The reaction of hemoglobin and oxygen is conventionally written

$$\underset{\text{Deoxyhemoglobin}}{Hb} + O_2 \rightleftharpoons \underset{\text{Oxyhemoglobin}}{HbO_2}$$

HEMOGLOBIN AND THE PHYSIOLOGICAL IMPLICATIONS OF THE OXYHEMOGLOBIN DISSOCIATION CURVE

The equilibrium point of the reversible reaction of hemoglobin and oxygen is, of course, dependent on how much oxygen the hemoglobin in blood is exposed to. This corresponds directly to the partial pressure of oxygen in the plasma under the conditions in the body. Thus, the P_{O_2} of the plasma *determines* the amount of oxygen that binds to the hemoglobin in the erythrocytes.

The Oxyhemoglobin Dissociation Curve

One way to express the proportion of hemoglobin that is bound to oxygen is as *percent saturation*. This is equal to the amount of oxygen in the blood (minus that part physically dissolved) divided by the oxygen-carrying capacity of the hemoglobin in the blood times 100%:

$$\% \text{ Hb saturation} = \frac{\text{blood } O_2 \text{ content}}{O_2 \text{ capacity of Hb}} \times 100\%$$

Note that the oxygen-carrying *capacity* of an individual is dependent on the amount of hemoglobin in that person's blood. The blood oxygen *content* is also dependent on the amount of hemoglobin present (as well as on the Po_2). Both content and capacity are expressed as milliliters of oxygen per 100 ml of blood. The percent hemoglobin saturation, on the other hand, expresses only a percentage and not an amount or volume of oxygen. Therefore, "percent saturation" is not interchangeable with "oxygen content." For example, two patients might have the same percent hemoglobin saturation, but if one has a lower blood hemoglobin concentration because of anemia, he or she will have a lower blood oxygen content.

The relationship between the Po_2 of the plasma and the percent of hemoglobin saturation is demonstrated graphically as the *oxyhemoglobin dissociation curve*. An oxyhemoglobin dissociation curve for normal blood is shown in Fig. 7-1.

The oxyhemoglobin dissociation curve is really a plot of how the availability of one of the reactants, oxygen (expressed as the Po_2 of the plasma), affects the reversible chemical reaction of oxygen and hemoglobin. The product, oxyhemoglobin, is expressed as percent saturation—really a percent of maximum for any given amount of hemoglobin.

As can be seen in Fig. 7-1, the relationship between Po_2 and HbO_2 is not linear; it is an "S-shaped" curve, steep at lower Po_2's and nearly flat when the Po_2 is above 70 torr. It is this S shape that is responsible for

Figure 7-1 A typical "normal" adult oxyhemoglobin dissociation curve for blood at a pH of 7.40, a Pco_2 of 40 torr and at 37°C. The P_{50} is the partial pressure of oxygen at which hemoglobin is 50% saturated with oxygen.

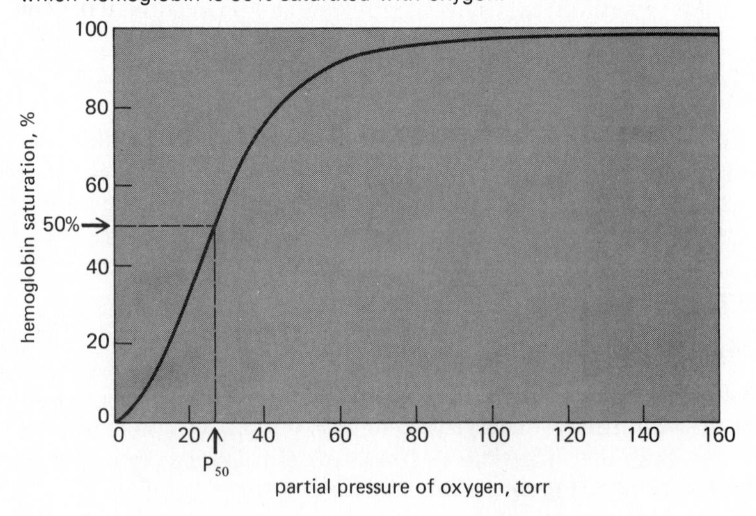

several very important physiological properties of the reaction of oxygen and hemoglobin. The reason that the curve is S-shaped and not linear is that it is actually a plot of four reactions rather than one. That is, as already stated, each of the four subunits of hemoglobin can combine with one molecule of oxygen. Indeed, it may be more correct to write the following equation:

$$Hb_4 + 4O_2 \rightleftharpoons Hb_4O_8$$

The reactions of the four subunits of hemoglobin with oxygen do not appear to occur simultaneously. Instead they are believed to occur sequentially in four steps, with an interaction between the subunits occurring in such a way that during the successive combinations of the subunits with oxygen, each combination facilitates the next. Similarly, dissociation of oxygen from hemoglobin subunits facilitates further dissociations. The dissociation curve for a single monomer of hemoglobin is far different from that for the tetramer, as seen in Fig. 7-4.

As already stated, for hemoglobin to participate in the transport of oxygen from the lungs to the tissues, it must combine with oxygen in the pulmonary capillaries and then release oxygen to the metabolizing tissues in the systemic capillaries. The oxyhemoglobin dissociation curve shown in Fig. 7-1 shows how this is accomplished.

Loading Oxygen in the Lung Mixed venous blood entering the pulmonary capillaries normally has a P_{O_2} of about 40 torr, as discussed in Chap. 5. At a P_{O_2} of 40 torr hemoglobin is about 75 percent saturated with oxygen, as seen in Fig. 7-1. Assuming a blood hemoglobin concentration of 15 g of hemoglobin per 100 ml of blood, this corresponds to 15.08 ml of oxygen per 100 ml of blood *bound to hemoglobin* plus an additional 0.12 ml of oxygen per 100 ml of blood *physically dissolved,* or a *total oxygen content* of approximately 15.20 ml of oxygen per 100 ml of blood: Oxygen-carrying *capacity* is

$$\frac{15\text{ g Hb}}{100\text{ ml blood}} \times \frac{1.34\text{ ml } O_2}{\text{gram Hb}} = \frac{20.1\text{ ml } O_2}{100\text{ ml blood}}$$

Oxygen *bound to hemoglobin* at a P_{O_2} of 40 torr (at 37°C and a pH of 7.4) is

$$\underset{\text{Capacity}}{\frac{20.1\text{ ml } O_2}{100\text{ ml blood}}} \times \underset{\%\text{ of saturation}}{75\%} = \underset{\text{Content}}{\frac{15.08\text{ ml } O_2}{100\text{ ml blood}}}$$

Oxygen physically dissolved at a P_{O_2} of 40 torr is

$$\frac{0.003 \text{ ml } O_2}{100 \text{ ml blood} \times P_{O_2} \text{ in torr}} \times 40 \text{ torr} = \frac{0.12 \text{ ml } O_2}{100 \text{ ml blood}}$$

Total blood oxygen content at a P_{O_2} of 40 torr (37°C and a pH of 7.4) is

$$\underset{\text{Bound to Hb}}{\frac{15.08 \text{ ml } O_2}{100 \text{ ml blood}}} + \underset{\text{Physically dissolved}}{\frac{0.12 \text{ ml } O_2}{100 \text{ ml blood}}} = \underset{\text{Total}}{\frac{15.20 \text{ ml } O_2}{100 \text{ ml blood}}}$$

As the blood passes through the pulmonary capillaries, it equilibrates with the alveolar P_{O_2} of about 100 torr. At a P_{O_2} of 100 torr hemoglobin is about 97.4 percent saturated with oxygen, as seen in Fig. 7-1. This corresponds to 19.58 ml oxygen per 100 ml of blood *bound to hemoglobin* plus 0.3 ml of oxygen per 100 ml of blood *physically dissolved,* or a *total oxygen content* of 19.88 ml of oxygen per 100 ml of blood: Oxygen *bound to hemoglobin* at a P_{O_2} of 100 torr (at 37°C and a pH of 7.4) is

$$\underset{\text{Capacity}}{\frac{20.1 \text{ ml } O_2}{100 \text{ ml blood}}} \times \underset{\text{\% of saturation}}{97.4\%} = \underset{\text{Content}}{\frac{19.58 \text{ ml } O_2}{100 \text{ ml blood}}}$$

Oxygen *physically dissolved* at a P_{O_2} of 100 torr is

$$\frac{0.003 \text{ ml } O_2}{100 \text{ ml blood} \times P_{O_2} \text{ in torr}} \times 100 \text{ torr} = \frac{0.3 \text{ ml } O_2}{100 \text{ ml blood}}$$

Total blood oxygen content at a P_{O_2} of 100 torr (37°C and a pH of 7.4) is

$$\underset{\text{Bound to Hb}}{\frac{19.58 \text{ ml } O_2}{100 \text{ ml blood}}} + \underset{\text{Physically dissolved}}{\frac{0.3 \text{ ml } O_2}{100 \text{ ml blood}}} = \underset{\text{Total}}{\frac{19.88 \text{ ml } O_2}{100 \text{ ml blood}}}$$

Thus, in passing through the lungs, each 100 ml of blood has loaded (19.88 − 15.20) ml O_2, or 4.68 ml of oxygen. Assuming a cardiac output of 5 liters/min, this means that approximately 234 ml O_2 is loaded into the blood per minute:

$$\frac{5 \text{ liters blood}}{\text{min}} \times \frac{46.8 \text{ ml } O_2}{\text{liter blood}} = \frac{234 \text{ ml } O_2}{\text{min}}$$

Note that the oxyhemoglobin dissociation curve is relatively flat at P_{O_2}'s greater than approximately 70 torr. This is very important physiologically because it means that there is only a small decrease in the *oxygen content* of blood equilibrated with a P_{O_2} of 70 torr instead of 100 torr. In fact, the curve shows that at a Po_2 of 70 torr, hemoglobin is still approximately 94.1 percent saturated with oxygen. This constitutes an important safety factor because a patient with a relatively low alveolar or arterial Po_2 of 70 torr (due to hypoventilation or intrapulmonary shunting, for example) is still able to load oxygen into the blood with little difficulty. A quick calculation shows that at 70 torr the total blood oxygen content is approximately 19.12 ml O_2 per 100 ml blood compared to the 19.88 ml O_2 per 100 ml blood at a Po_2 of 100 torr. These calculations show that Po_2 is often a more sensitive diagnostic indicator of the status of a patient's respiratory system than is the arterial oxygen content. Of course, the oxygen content is more important physiologically to the patient.

It should also be noted that since hemoglobin is approximately 97.4 percent saturated at a Po_2 of 100 torr, raising the alveolar Po_2 above 100 torr can load little additional oxygen onto hemoglobin (only about 0.52 ml O_2 per 100 ml blood at a hemoglobin concentration of 15 g per 100 ml blood). Hemoglobin is fully saturated with oxygen at a Po_2 of about 250 torr.

Unloading Oxygen at the Tissues As blood passes from the arteries into the systemic capillaries, it is exposed to lower Po_2's and oxygen is released by the hemoglobin. The Po_2 in the capillaries varies from tissue to tissue, being very low in some (e.g., myocardium) and relatively higher in others (e.g., kidney). As can be seen in Fig. 7-1, the oxyhemoglobin dissociation curve is very steep in the range of 40 to 10 torr. This means that small decreases in Po_2 can result in a substantial further dissociation of oxygen and hemoglobin, unloading more oxygen for use by the tissues. At a Po_2 of 40 we saw that hemoglobin is about 75 percent saturated with oxygen, with a total blood oxygen content of 15.2 ml O_2 per 100 ml blood. At a Po_2 of 20 torr, hemoglobin is only 32 percent saturated with oxygen. The total blood oxygen content is only 6.49 ml O_2 per 100 ml of blood, a decrease of 8.71 ml O_2 per 100 ml blood for only a 20-torr decrease in Po_2.

The unloading of oxygen at the tissues is also facilitated by other physiological factors that can *alter the shape and position* of the oxyhemoglobin dissociation curve. These include the pH, Pco_2, temperature of the blood, and concentration of 2,3-diphosphoglycerate (2,3-DPG) in the erythrocytes.

INFLUENCES ON THE OXYHEMOGLOBIN DISSOCIATION CURVE

Figure 7-2 shows the influence of alterations in temperature, pH, Pco_2, and 2,3-DPG on the oxyhemoglobin dissociation curve. High temperature, low pH, high Pco_2, and elevated levels of 2,3-DPG all "shift the oxyhemoglobin dissociation curve to the right." That is, for any particular

Effects of pH, PCO2 + temp are more pronounced @ lower PO2's than higher PO2's. → ↑ effect in enhancing unloading of O2 @ tissues than they interfere c̄ its loading @ lungs.

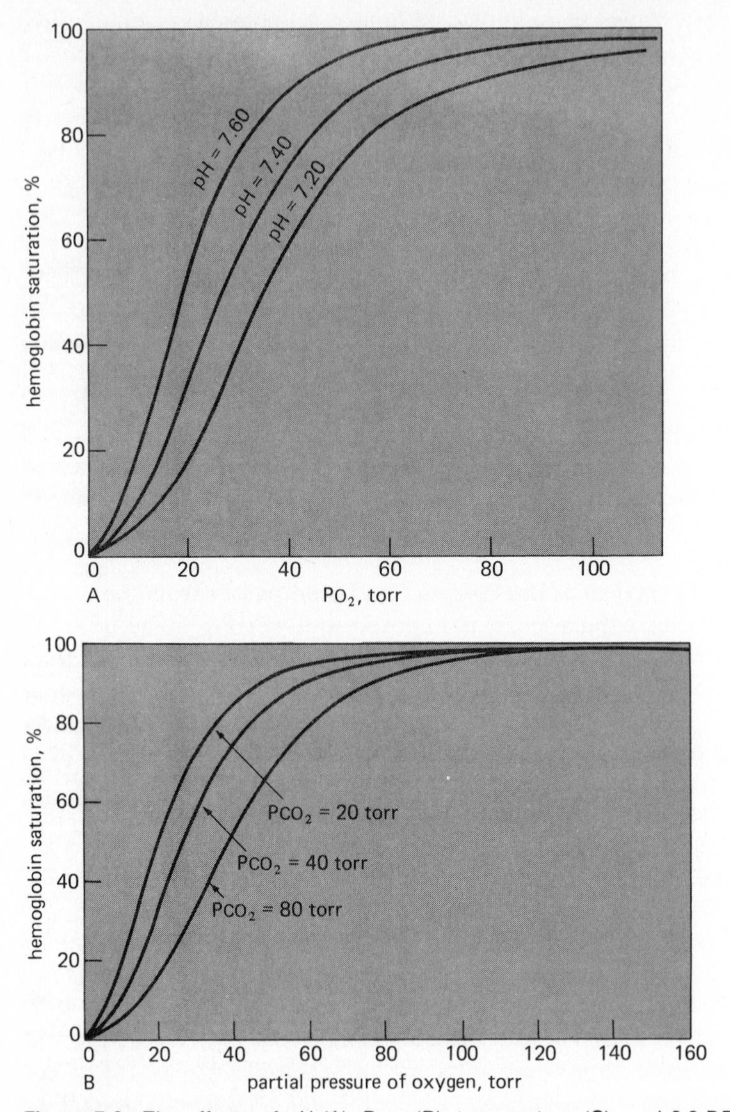

Figure 7-2 The effects of pH (A), P_{CO_2} (B), temperature (C), and 2,3-DPG (D) on the oxyhemoglobin dissociation curve.

P_{O_2} there is less oxygen chemically combined with hemoglobin at higher temperatures, lower pH's, higher P_{CO_2}'s, and elevated levels of 2,3-DPG.

Effects of pH and P_{CO_2}

The effects of blood pH and P_{CO_2} on the oxyhemoglobin dissociation curve are shown in Figs. 7-2*A* and *B*. Low pH's or high P_{CO_2}'s both shift the curve to the right. High pH's and low P_{CO_2}'s both shift the curve to the

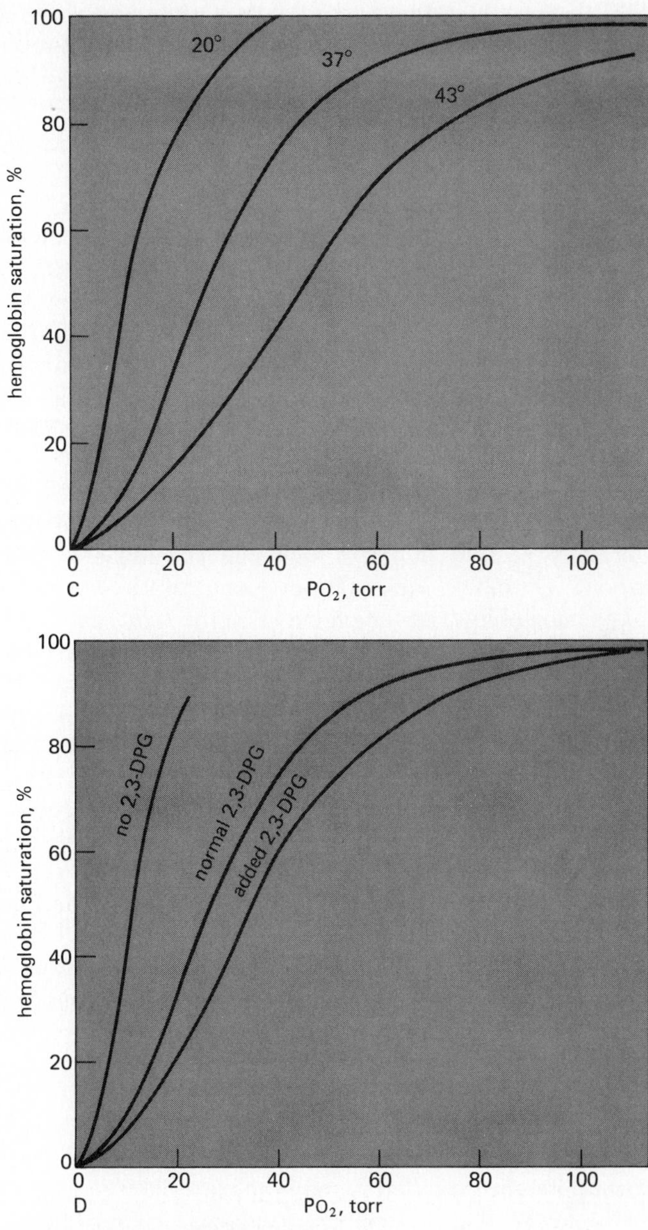

Figure 7-2 (*Continued*)

left. Because high P_{CO_2}'s in blood are often associated with low pH's, as will be discussed later in this chapter and in Chap. 8, these two effects may often occur together. The influence of pH (and P_{CO_2}) on the oxyhemoglobin dissociation curve is referred to as the *Bohr effect*. The Bohr effect will be discussed in greater detail at the end of this chapter.

Effects of Temperature

Figure 7-2*C* shows the effects of blood temperature on the oxyhemoglobin dissociation curve. High temperatures shift the curve to the right; low temperatures shift the curve to the left. At very low blood temperatures hemoglobin has such a high affinity for oxygen that it does not release it even at very low P_{O_2}'s. It should also be noted that oxygen is more soluble in water or plasma at lower temperatures than it is at normal body temperature. At 20°C about 50 percent more oxygen will dissolve in plasma.

Effects of 2,3-DPG

2,3-DPG is produced by erythrocytes during anaerobic glycolysis and is normally present in fairly high concentrations within red blood cells (about 15 mol/g Hb). Higher concentrations of 2,3-DPG shift the oxyhemoglobin dissociation curve to the right, as shown in Fig. 7-2*D*. It has been demonstrated that more 2,3-DPG is produced during chronic hypoxic conditions, thus shifting the dissociation curve to the right and allowing more oxygen to be released from hemoglobin at a particular P_{O_2}. Very low levels of 2,3-DPG shift the curve far to the left, as shown in the figure. This means that blood deficient in 2,3-DPG does not unload much oxygen except at very low P_{O_2}'s. It is important to note that blood stored in blood banks for as little as 1 week has been consistently shown to have very low levels of 2,3-DPG. Delivery of banked blood to patients may result in greatly decreased oxygen unloading to the tissues, unless steps are taken to restore the normal levels of 2,3-DPG. In summary, 2,3-DPG is an important regulating mechanism in the release of oxygen by hemoglobin. Without its presence, hemoglobin's high affinity for oxygen would impair the oxygen supply of the tissues.

Physiological Consequences of the Effects of Temperature, pH, P_{CO_2}, and 2,3-DPG

As blood enters metabolically active tissues, it is exposed to an environment different from that found in the arterial tree. The P_{CO_2} is higher, the pH is lower, and the temperature is also higher than that of the arterial blood. It is evident, then, that in our original discussion of the oxyhemoglobin dissociation curve shown in Fig. 7-1 we were neglecting some important factors. The curve shown in Fig. 7-1 is for blood at 37°C, with a pH of 7.4, and a P_{CO_2} of 40 torr. Blood in metabolically active tissues and *venous blood* are not under these conditions because they have been exposed to a different environment. Since low pH, high P_{CO_2}, and higher temperature *all shift the oxyhemoglobin dissociation curve to the right,* they all can help to unload oxygen from hemoglobin at the tissues. On the other hand, as the venous blood returns to the lung and CO_2 leaves the blood (which increases the pH), the affinity of hemoglobin for oxygen increases as the curve shifts back to the left, as shown in Fig. 7-3.

Note that the effects of pH, P_{CO_2}, and temperature shown in Fig. 7-2 are all more pronounced at lower P_{O_2}'s than at the higher P_{O_2}'s. That is, they have a more profound effect on enhancing the unloading of oxygen at the tissues than they interfere with its loading at the lungs.

A convenient way to discuss shifts in the oxyhemoglobin dissociation curve is the P_{50}, shown in Figs. 7-1 and 7-3. The P_{50} is the P_{O_2} at which 50 percent of the hemoglobin present in the blood is in the deoxyhemoglobin state and 50 percent is in the oxyhemoglobin state. At a temperature of 37°C, a pH of 7.4, and a P_{CO_2} of 40 torr, normal human blood has a P_{50} of 26 or 27 torr. If the oxyhemoglobin dissociation curve is shifted to the right, the P_{50} increases. If it is shifted to the left, the P_{50} decreases.

Other Factors Affecting Oxygen Transport

Anemia Most forms of anemia do not affect the oxyhemoglobin dissociation curve if the association of oxygen and hemoglobin are expressed

Figure 7-3 Oxyhemoglobin dissociation curves for arterial and venous blood. The venous curve is shifted to the right because the pH is lower and the P_{CO_2} (and possibly temperature) is higher. The rightward shift results in a higher P_{50} for venous blood. a = arterial point (P_{O_2} = 100 torr); $\bar{v}$ = mixed venous point (P_{O_2} = 40 torr).

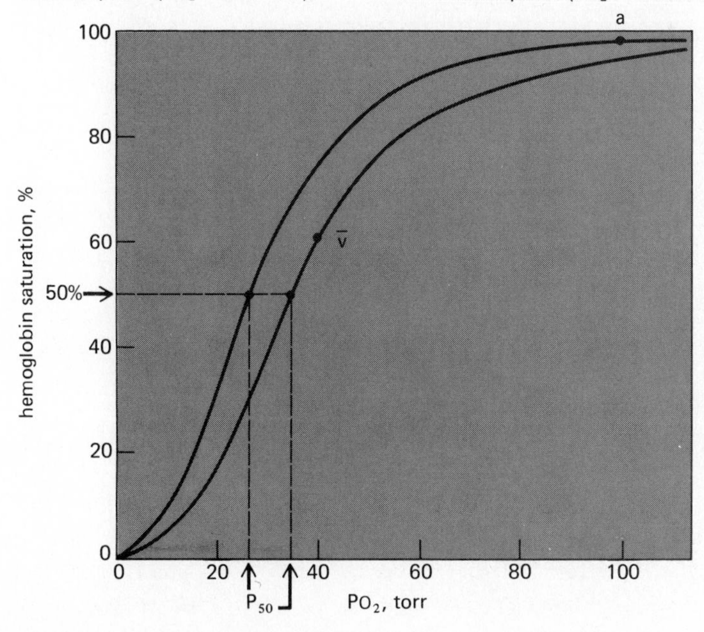

as percent saturation. For example, anemia secondary to blood loss does not affect the combination of oxygen and hemoglobin for the remaining erythrocytes. It is the *amount* of hemoglobin that decreases, not the percent saturation or even the arterial P_{O_2}. The arterial *content* of oxygen, however, in milliliters of oxygen per 100 ml blood, is reduced because of the decreased amount of hemoglobin per 100 ml of blood, as seen in Fig. 7-4*A*.

Carbon Monoxide Carbon monoxide has a much greater affinity for hemoglobin than does oxygen, as discussed in Chap. 6. It can therefore effectively block the combination of oxygen with hemoglobin because oxygen cannot be bound to iron atoms already combined with carbon monoxide. Carbon monoxide has a second deleterious effect: It shifts the oxyhemoglobin dissociation curve to the left. Thus carbon monoxide can both prevent the loading of oxygen into the blood in the lungs and it can also interfere with the unloading of oxygen at the tissues. This can be seen in Fig. 7-4*A*.

Carbon monoxide is particularly dangerous for several reasons. A person breathing very low concentrations of carbon monoxide can slowly

Figure 7-4 *A*. The effects of carbon monoxide and anemia on the carriage of oxygen by hemoglobin. Note that the ordinate is expressed as the *volume* of oxygen bound to hemoglobin in milliliters of O_2 per 100 ml blood. *B*. A comparison of the oxyhemoglobin dissociation curves for normal adult hemoglobin (HbA) and fetal hemoglobin (HbF). *C*. Dissociation curves for normal adult hemoglobin (HbA), a single monomeric subunit of hemoglobin (Hb subunit), and myoglobin (Mb).

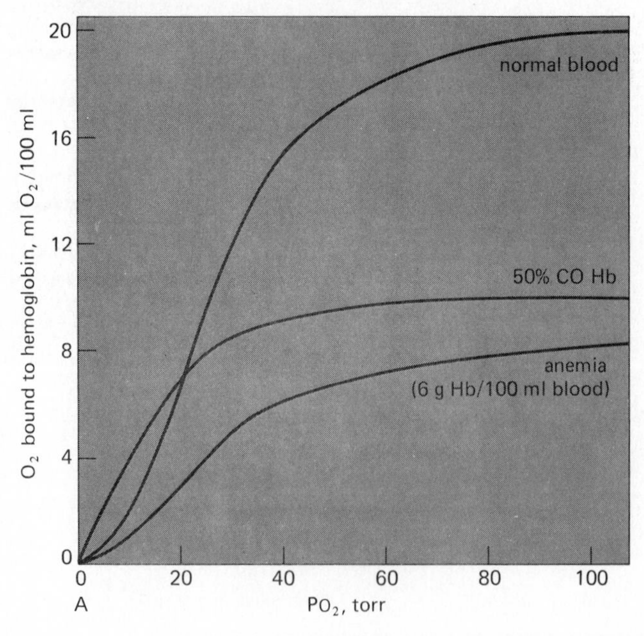

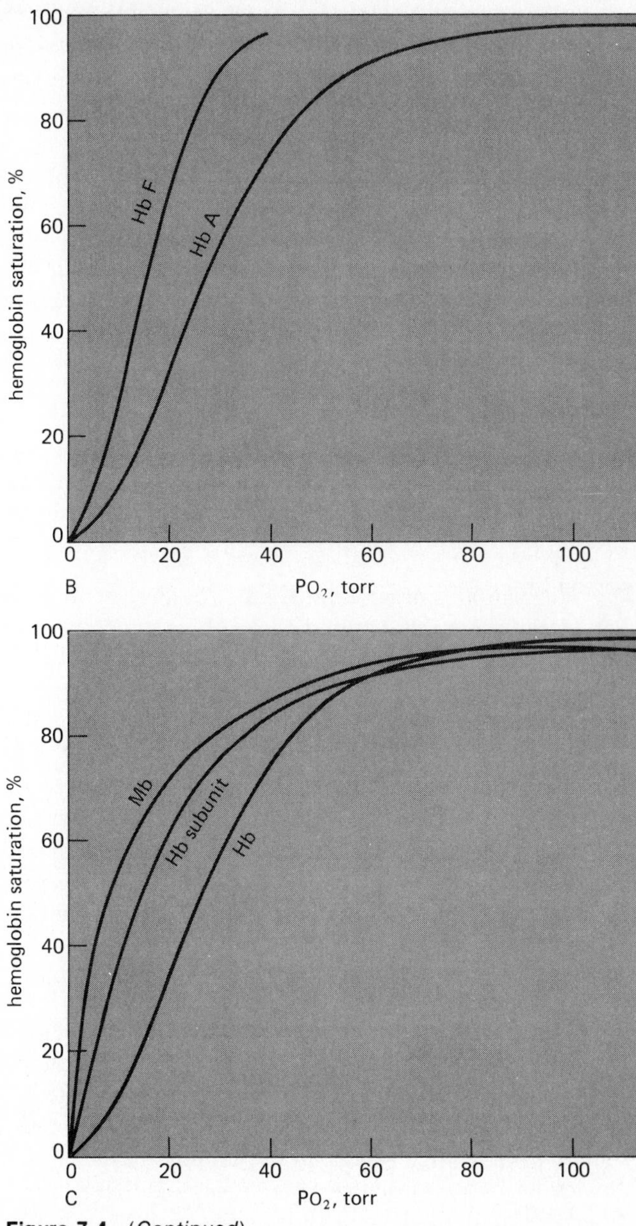

Figure 7-4 (*Continued*)

reach life-threatening levels of carboxyhemoglobin (COHb) in the blood because carbon monoxide has such a high affinity for hemoglobin. The effect is cumulative. What is worse is that a person breathing carbon monoxide does not know he or she is doing so—the gas is colorless, odorless, and tasteless and does not elicit any reflex coughing or sneezing, increase in ventilation, or feeling of difficulty in breathing.

Smoking and living in urban areas cause small amounts of carboxyhemoglobin to be present in the blood of normal healthy adults. A nonsmoker who lives in a rural area may have only about 1 percent carboxyhemoglobin; an urban person who is a heavy smoker may have 5 to 8 percent carboxyhemoglobin in his or her blood.

Methemoglobin Methemoglobin is hemoglobin with iron in the ferric (Fe^{3+}) state. It can be caused by nitrite poisoning and in toxic reactions to oxidant drugs or it is found congenitally in patients with hemoglobin M. Iron atoms in the Fe^{3+} state will not combine with oxygen.

Hemoglobins other than HbA As already discussed in this chapter, variants of the normal adult hemoglobin A may have different affinities for oxygen. Fetal hemoglobin (HbF) in red blood cells has a dissociation curve to the left of that for HbA, as shown in Fig. 7-4*B*. This is perfectly reasonable because fetal P_{O_2}'s are much lower than those of an adult: The curve is located properly for its operating range. The shape of the HbF curve in blood appears to be due to the fact that 2,3-DPG has little effect on the affinity of HbF for oxygen. Indeed, the curve is similar to that of HbA without 2,3-DPG (Fig. 7-2*D*). Abnormal hemoglobins may have either increased or decreased affinities for oxygen. For example, Hb Seattle and Hb Kansas have lower affinities for oxygen than does HbA; Hb Rainier has a higher affinity for oxygen.

Myoglobin Myoglobin (Mb), a heme protein that occurs naturally in muscle cells, consists of a single polypeptide chain attached to a heme group. It can therefore combine chemically with a single molecule of oxygen and is similar structurally to a single subunit of hemoglobin. As can be seen in Fig. 7-4*C*, the hyperbolic dissociation curve of myoglobin (which is similar to that of a single hemoglobin subunit) is far to the left of that of normal adult hemoglobin. That is, at lower P_{O_2}'s much more oxygen remains bound to myoglobin. Myoglobin can therefore act to store and transport oxygen in skeletal muscle. As blood passes through the muscle, oxygen leaves hemoglobin and binds to myoglobin. It can be released from the myoglobin when conditions cause lower P_{O_2}'s.

Cyanosis Cyanosis is not really an influence on the transport of oxygen but rather is a sign of poor transport of oxygen. Cyanosis occurs when more than 5 g of hemoglobin per 100 ml of arterial blood is in the deoxy state. It is a bluish purple discoloration of the skin, nail beds, and mucous membranes and its presence is indicative of an abnormally high concentration of deoxyhemoglobin in the arterial blood. Its absence, however, does not exclude hypoxemia because an anemic patient with hypoxemia may not have sufficient hemoglobin to appear cyanotic. Patients with abnormally high levels of hemoglobin in their arterial blood, such as those with polycythemia, may appear cyanotic without being hypoxemic.

TRANSPORT OF CARBON DIOXIDE BY THE BLOOD

Carbon dioxide is carried in the blood in physical solution, chemically combined to amino acids in blood proteins, and as bicarbonate ions. About 200 to 250 ml of carbon dioxide is produced by the tissue metabolism of a resting 70-kg person each minute and must be carried by the venous blood to the lung for removal from the body. At a cardiac output of 5 liters/min, each 100 ml of blood passing through the lungs must therefore unload 4 to 5 ml of carbon dioxide.

Physically Dissolved

Carbon dioxide is about 20 times as soluble in the plasma (and inside the erythrocytes) as is oxygen. About 5 to 10 percent of the total carbon dioxide transported by the blood is carried in physical solution.

About 0.0006 ml of CO_2 per torr of P_{CO_2} will dissolve in 1 ml of plasma at 37°C. One hundred ml of plasma or whole blood at a P_{CO_2} of 40 torr therefore contains about 2.4 ml of CO_2 in *physical solution.* As can be seen in the carbon dioxide dissociation curve for whole blood shown in Fig. 7-5, at 40 torr the total CO_2 content of whole blood is about 48 ml of CO_2 per 100 ml of blood, and so approximately 5 percent of the carbon dioxide carried in the arterial blood is in physical solution. Similarly, multiplying 0.06 ml of CO_2 per 100 ml of blood per torr of P_{CO_2} times a venous P_{CO_2} of 45 torr shows that about 2.7 ml of CO_2 is physically dissolved in the mixed venous blood. The total carbon dioxide content of venous blood is about 52.5 ml of CO_2 per 100 ml of blood, and so a little more than 5 percent of the total carbon dioxide content of venous blood is in physical solution.

As Carbamino Compounds

Carbon dioxide can combine chemically with the terminal amine groups in blood proteins, forming *carbamino compounds.*

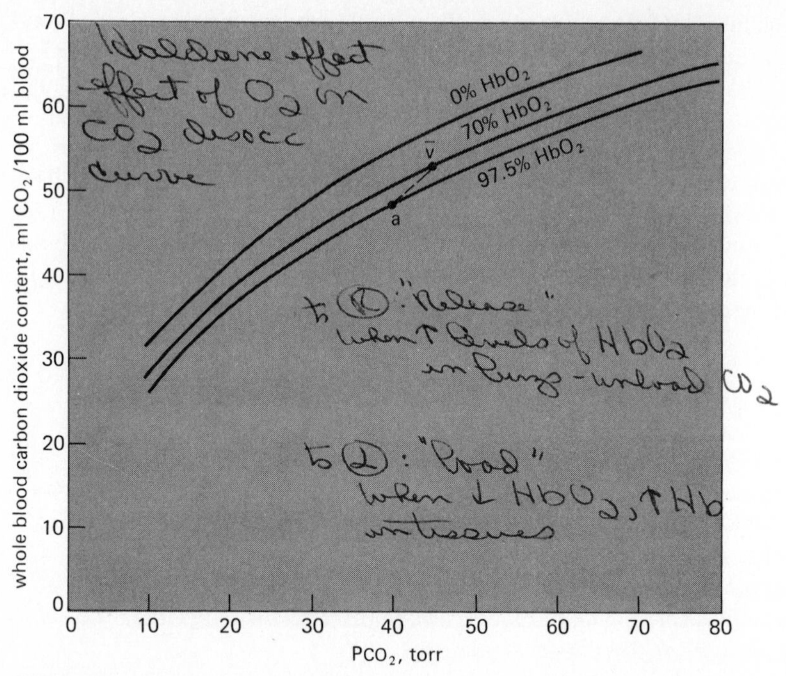

Figure 7-5 Carbon dioxide dissociation curves for whole blood (37°C) at different hemoglobin oxygen saturations. Note that the ordinate is whole-blood CO_2 content in milliliters of CO_2 per 100 ml of blood.

$$\underset{\text{Terminal amine group}}{R{-}N\begin{smallmatrix} H \\ H \end{smallmatrix}} + CO_2 \rightleftharpoons \underset{\text{Carbamino compound}}{R{-}N\begin{smallmatrix} H \\ COO^- \end{smallmatrix}} + H^+$$

The reaction occurs rapidly; no enzymes are necessary.

Because the protein found in greatest concentration in the blood is the globin of hemoglobin, most of the carbon dioxide transported in this manner is bound to amino acids of hemoglobin. Deoxyhemoglobin can bind more carbon dioxide as carbamino groups than can oxyhemoglobin. Therefore, as the hemoglobin in the venous blood enters the lung and combines with oxygen, it releases carbon dioxide from its terminal amine groups. About 5 to 10 percent of the total carbon dioxide content of blood is in the form of carbamino compounds.

As Bicarbonate

The remaining 70 to 90 percent of the carbon dioxide transported by the blood is carried as bicarbonate ions. This is made possible by the following reaction:

$$CO_2 + H_2O \underset{}{\overset{\text{carbonic anhydrase}}{\rightleftharpoons}} H_2CO_3 \rightleftharpoons H^+ + HCO_3^-$$

Carbon dioxide can combine with water to form carbonic acid, which then dissociates into a hydrogen ion and a bicarbonate ion.

Very little carbonic acid is formed by the association of water and carbon dioxide without the presence of the enzyme carbonic anhydrase because the reaction occurs so slowly. Carbonic anhydrase, which is present in high concentration in erythrocytes (but not in plasma), makes the reaction proceed about 13,000 times faster. Hemoglobin also plays an integral role in the transport of carbon dioxide because it can accept the hydrogen ion liberated by the dissociation of carbonic acid, thus allowing the reaction to continue. This will be discussed in detail in the last section of this chapter.

THE CARBON DIOXIDE DISSOCIATION CURVE

The carbon dioxide dissociation curve for *whole blood* is shown in Fig. 7-5. Within the normal physiological range of P_{CO_2}'s the curve is nearly a straight line, with no steep and flat portions. If it is plotted on axes similar to those for oxygen, the carbon dioxide dissociation curve for whole blood is steeper than the oxygen dissociation curve for whole blood. That is, there is a greater change in CO_2 content per torr change in P_{CO_2} than there is in O_2 content per torr change in P_{O_2}.

The carbon dioxide dissociation curve for whole blood is shifted to the right at greater levels of oxyhemoglobin and shifted to the left at greater levels of deoxyhemoglobin. This is known as the *Haldane effect,* which will be explained in the next section. The Haldane effect allows the blood to load more carbon dioxide at the tissues, where there is more deoxyhemoglobin, and unload more carbon dioxide in the lungs, where there is more oxyhemoglobin.

THE BOHR AND HALDANE EFFECTS EXPLAINED

The Bohr and Haldane effects are both explained by the fact that *deoxyhemoglobin is a weaker acid than oxyhemoglobin.* That is, deoxyhemoglobin more readily accepts the hydrogen ion liberated by the dissociation of carbonic acid, thus permitting more carbon dioxide to be

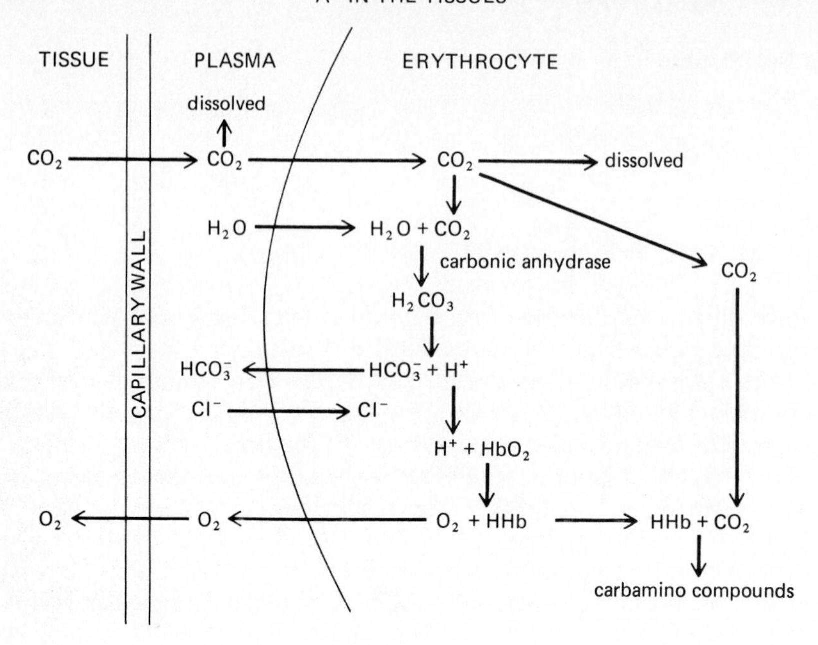

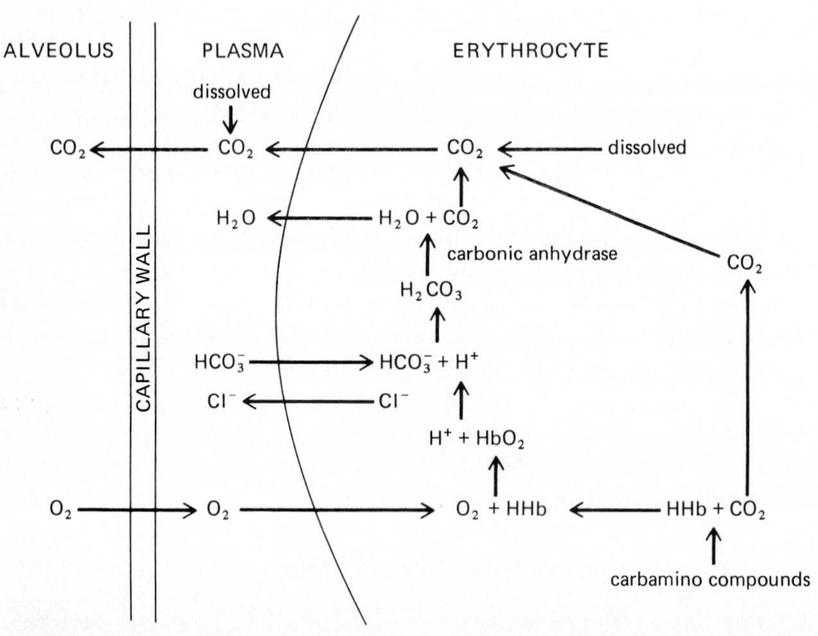

Figure 7-6 Schematic representation of uptake and release of carbon dioxide and oxygen at the tissues (*A*) and in the lung (*B*). Note that negligible amounts of carbon dioxide can form carbamino compounds with blood proteins other than hemoglobin and may also be hydrated in trivial amounts in the plasma to form carbonic acid and then bicarbonate (not shown in diagram).

transported in the form of bicarbonate ion. This is referred to as the *isohydric shift*. Conversely, the association of hydrogen ions with the amino acids of hemoglobin lowers the affinity of hemoglobin for oxygen, thus shifting the oxyhemoglobin dissociation curve to the right at low pH's or high P_{CO_2}'s. The following "equation" can therefore be written:

$$H^+Hb + O_2 \rightleftharpoons H^+ + HbO_2$$

These effects can be seen in the schematic diagrams of oxygen and carbon dioxide transport shown in Fig. 7-6.

At the tissues the P_{O_2} is low and the P_{CO_2} is high. Carbon dioxide dissolves in the plasma and some diffuses into the erythrocyte. Some of this carbon dioxide dissolves in the cytosol, some forms carbamino compounds with hemoglobin, and some is hydrated by carbonic anhydrase to form carbonic acid. At low P_{O_2}'s, there are substantial amounts of deoxyhemoglobin in the erythrocytes and the deoxyhemoglobin is able to accept the hydrogen ion liberated by the dissociation of carbonic acid. Bicarbonate ions diffuse out of the erythrocyte through the cell membrane much more readily than do hydrogen ions. Because more bicarbonate ions than hydrogen ions diffuse out of the erythrocyte, electrical neutrality is maintained by the diffusion of chloride ions into the cell, matching the movement of bicarbonate ions out of the cell. This is the "chloride shift." Small amounts of water also move into the cell to maintain the osmotic equilibrium.

At the lung the P_{O_2} is high and the P_{CO_2} is low. As oxygen combines with hemoglobin, the hydrogen ions that were taken up when it was in the deoxyhemoglobin state are released. They combine with bicarbonate ions, forming carbonic acid. This breaks down into carbon dioxide and water. At the same time carbon dioxide is also released from the carbamino compounds. Carbon dioxide then diffuses out of the red blood cell and plasma and into the alveoli. A chloride shift opposite in direction to that in the tissues also occurs to maintain electrical neutrality.

Chapter 8

The Regulation of Acid-Base Status

OBJECTIVES

The student understands the basic concepts of the regulation of the acid-base status of the body.

1 Defines acids, bases, and buffers.
2 Lists the buffer systems available in the human body.
3 Describes the interrelationships of the pH, the P_{CO_2} of the blood, and the plasma bicarbonate concentration and states the Henderson-Hasselbalch equation.
4 States the normal ranges of arterial pH, P_{CO_2}, and bicarbonate concentration and defines alkalosis and acidosis.
5 Lists the potential causes of respiratory acidosis and alkalosis and metabolic acidosis and alkalosis.
6 Discusses the respiratory and renal mechanisms that help to compensate for acidosis and alkalosis.
7 Evaluates blood gas data to determine a subject's acid-base status.

The maintenance of a relatively constant internal environment is one of the major physiological functions of the organ systems of the body. Body temperature, fluid volume and osmolarity, and electrolytes, including acids and bases, are normally carefully regulated. A thorough knowledge of the mechanisms that control these variables is essential to clinical practice.

The respiratory system is intimately involved in the maintenance of the balance of acids and bases in the body. This chapter will introduce the major concepts of acid-base balance, particularly with respect to the respiratory system; a more detailed study of this important subject is strongly encouraged.

THE CHEMISTRY OF ACIDS, BASES, AND BUFFERS

Although there are several ways to define acids and bases, the most useful physiologically is to define an *acid* as a substance that can donate a hydrogen ion (a proton) to another substance, and a *base* as a substance that can accept a hydrogen ion from another substance. A *strong acid* is a substance that is completely or almost completely dissociated into a hydrogen ion and its corresponding or *conjugate* base in dilute aqueous solution; a *weak acid* is only slightly ionized in aqueous solution. In general, a strong acid has a weak conjugate base and a weak acid has a strong conjugate base. The strength of an acid or a base should not be confused with its concentration.

A *buffer* is a mixture of substances in aqueous solution, usually a combination of a weak acid and its conjugate base, which can resist changes in hydrogen ion concentration when strong acids or bases are added. That is, the changes in hydrogen ion concentration that occur when a strong acid or base is added to a buffer system are much smaller than those that would occur if the same amount of acid or base were added to pure water or other nonbuffer solution.

The Quantification of Acidity

The acidity of a solution is determined by the *activity* of the hydrogen ions in the solution. The hydrogen ion activity, which is denoted by the symbol α_{H^+}, is closely related to the concentration of hydrogen ions ($[H^+]$) in a solution. In extremely dilute solutions the hydrogen ion activity is equal to the hydrogen ion concentration; in highly concentrated solutions the activity is less than the concentration. The hydrogen ion concentration of blood is low enough that the hydrogen ion activity may be considered to be equal to the hydrogen ion concentration.

The hydrogen ion activity of pure water is about 1.0×10^{-7} mol/liter. By convention, solutions with hydrogen ion activities above 10^{-7} mol/liter are considered to be acid; those with hydrogen ion activities below 10^{-7}

are considered to be alkaline. The range of hydrogen ion concentrations or activities in the *body* is normally from about 10^{-1} for gastric acid to about $10^{-7.5}$ for the most alkaline pancreatic secretion. This wide range of hydrogen ion activities necessitates the use of the more convenient *pH scale*. The pH of a solution is the negative logarithm of its hydrogen ion activity. With the exception of the highly concentrated gastric acid, in most instances in the body the hydrogen ion activity is about equal to the hydrogen ion concentration. Therefore,

$$\mathrm{pH} = -\log(\alpha_{\mathrm{H}^+})$$

or $$\mathrm{pH} = -\log[\mathrm{H}^+]$$

Thus, the pH of gastric acid is on the order of 1; the pH of the alkaline pancreatic secretion may be as high as 7.5.

The pH of *arterial blood* is normally close to 7.40, with a normal range considered to be about 7.35 to 7.45. Under pathological conditions the extremes of arterial blood pH have been noted to range as high as 7.8 and as low as 6.9. These correspond to hydrogen ion concentrations as follows [hydrogen ion concentrations are expressed as *nanomoles* (10^{-9} mol/liter) for convenience]:

pH 6.90 = 126 nmol/liter
pH 7.00 = 100 nmol/liter
pH 7.10 = 79 nmol/liter
pH 7.20 = 63 nmol/liter
pH 7.30 = 50 nmol/liter
pH 7.40 = 40 nmol/liter
pH 7.50 = 32 nmol/liter
pH 7.60 = 25 nmol/liter
pH 7.70 = 20 nmol/liter
pH 7.80 = 16 nmol/liter

Note that the pH scale is "inverted" by the negative sign and is also logarithmic as it is defined. Therefore an *increase* in pH from 7.40 to 7.70 represents a *decrease* in hydrogen ion concentration. In fact, the increase of only 0.3 pH units indicates that hydrogen ion concentration was cut in *half*.

The Importance of Body pH Regulation

Hydrogen ions are the most reactive cations in body fluids, and they interact with negatively charged regions of other molecules, such as those of body proteins. Interactions of hydrogen ions with negatively charged functional groups of proteins can lead to marked changes in protein structural conformations with resulting alterations in the behavior of the proteins. An example of this was already seen in Chap. 7, where hemoglobin was noted to combine with less oxygen at lower pH's (the Bohr effect). Alterations in the structural conformations and charges of protein enzymes will obviously affect their activities, with resulting alterations in the functions of body tissues. The absorption and efficacy of drugs administered by the physician may also be affected by the pH. Extreme changes in the hydrogen ion concentration of the body can result in loss of organ system function and structural integrity; pH's above approximately 7.80 or below 6.9 are not compatible with life.

Sources of Acids in the Body

Under normal circumstances the metabolism of the cells of the body is the main source of acids in the body. These acids are the waste products of substances ingested as foodstuffs. The greatest source of hydrogen ions is the carbon dioxide produced as one of the end products of the oxidation of glucose and fatty acids during aerobic metabolism. The hydration of carbon dioxide results in the formation of carbonic acid, which then can dissociate into a hydrogen ion and a bicarbonate ion, as discussed in Chap. 7. This process is reversed in the pulmonary capillaries, and CO_2 then diffuses through the alveolar-capillary barrier into the alveoli, from which it is removed by alveolar ventilation. Carbonic acid is therefore said to be a *volatile acid* because it can be converted into a gas and then removed from an open system like the body. Very great amounts of carbonic acid can be removed from the lungs by alveolar ventilation: Under normal circumstances about 24,000 meq of carbonic acid are removed via the lungs daily.

A much smaller quantity of *fixed,* or *nonvolatile, acids* is also normally produced during the course of the metabolism of foodstuffs. The fixed acids produced by the body include sulfuric acid, which originates from the oxidation of sulfur-containing amino acids such as cysteine; phosphoric acid from the oxidation of phospholipids and phosphoproteins; hydrochloric acid, which is produced during the conversion of ingested ammonium chloride to urea and by other reactions; and lactic acid from the anaerobic metabolism of glucose. Of course, lactic acid is sometimes converted to carbon dioxide, and so it is not always a fixed acid. Other fixed acids may be ingested accidentally or formed in abnormally large quantities by disease processes, such as the acetoacetic and butyric acid formed during diabetic ketoacidosis. About 50 meq of fixed acids are normally removed from the body each day. This removal is mainly accomplished by the kidneys, as will be discussed later in this chapter. Some may also be removed via the gastrointestinal tract. Fixed acids normally represent only about 0.2 percent of the total body acid production.

BUFFER SYSTEMS OF THE HUMAN BODY

The body contains a variety of substances that can act as buffers in the physiological pH range. These include bicarbonate, phosphate, and proteins. One way to express the ability of a substance to act as a buffer is its *buffer value.* The buffer value of a solution is the amount of hydrogen ions in milliequivalents per liter that can be added to or removed from the solution with a resultant change of one pH unit. Another way is to determine the substance's titration curve. Another important consideration is the pK of the acid.

An acid, HA, can dissociate into a hydrogen ion, H^+, and its base, A^-.

$$HA \rightleftharpoons H^+ + A^-$$

According to the *law of mass action,* the relationship between the undissociated acid and the proton and the base at equilibrium can be expressed as the following ratio:

$$\frac{[H^+][A^-]}{[HA]} = K$$

That is, the product of the concentrations of hydrogen ion and base divided by the concentration of the acid is equal to a constant, K, the *dissociation constant*. This can be rearranged to

$$[H^+] = K \frac{[HA]}{[A^-]}$$

After taking the logarithm of both sides,

$$\log [H^+] = \log K + \log \frac{[HA]}{[A^-]}$$

After multiplying both sides by -1,

$$-\log [H^+] = -\log K + \log \frac{[A^-]}{[HA]}$$

or

$$pH = pK + \log \frac{[A^-]}{[HA]}$$

This is the general form of the *Henderson-Hasselbalch equation.*

The buffer value or buffering capacity of a buffer pair is greatest at or near the pK of the weak acid. Note that when the concentrations of HA and A^- are equal, the pH of a solution is equal to its pK.

As already stated, the human body contains a number of buffers and buffer pairs. The *isohydric principle* states that all of the buffer pairs in a homogeneous solution are in equilibrium with the same hydrogen ion concentration. For this reason, all of the buffer pairs in the *plasma* behave similarly, with the relative concentrations of their undissociated acids and their bases determined by their respective pKs:

$$\mathrm{pH} = \mathrm{p}K_1 + \log \frac{[A_1^-]}{[HA_1]} = \mathrm{p}K_2 + \log \frac{[A_2^-]}{[HA_2]} = \mathrm{p}K_3 + \log \frac{[A_3^-]}{[HA_3]}$$

An implication of the isohydric principle is that the detailed analysis of a single buffer pair, like the bicarbonate buffer system, can reveal a great deal about the chemistry of all of the plasma buffers.

Bicarbonate

The bicarbonate buffer system consists of the buffer pair of the weak acid, carbonic acid, and its conjugate base, bicarbonate. As already stated, in the body

$$\underset{\text{Gas phase}}{CO_2} \rightleftharpoons \underset{\substack{\text{Dissolved}\\ \text{in the}\\ \text{aqueous phase}}}{CO_2} + H_2O \xrightleftharpoons{\text{carbonic anhydrase}} H_2CO_3 \rightleftharpoons H^+ + HCO_3^-$$

The ability of the bicarbonate system to function as a buffer of fixed acids in the body is largely due to the ability of the lungs to remove carbon dioxide from the body. In a closed system bicarbonate would not be nearly as effective.

At a temperature of 37°C about 0.03 mmol of carbon dioxide per torr of P_{CO_2} will dissolve in a liter of plasma. (Note that the solubility of CO_2 was expressed as *milliliters of CO_2* per *100 ml* of plasma in Chap. 7.) Therefore the carbon dioxide *dissolved* in the plasma, expressed as millimoles per liter, is equal to $0.03 \times P_{CO_2}$. At body temperature in the plasma, the equilibrium of the second part of the series of equations given above is far to the left so that there is roughly 1000 times as much carbon dioxide present physically dissolved in the plasma as there is in the form of carbonic acid. The dissolved carbon dioxide is in *equilibrium* with the carbonic acid, though, and so *both* the dissolved carbon dioxide and the carbonic acid are considered as the undissociated HA in the Henderson-Hasselbalch equation for the bicarbonate system:

$$pH = pK + \log \frac{[HCO_3^-]\ p}{[CO_2 + H_2CO_3]}$$

The concentration of carbonic acid is negligible, and so

$$pH = pK' + \log \frac{[HCO_3^-]\ p}{0.03 \times P_{CO_2}}$$

where pK' is the pK of the HCO_3^-/CO_2 system in blood.

The pK' of this system at physiological pH's and at 37°C is 6.1. Therefore, at an arterial pH of 7.40 and an arterial P_{CO_2} of 40 torr,

$$7.40 = 6.1 + \log \frac{[HCO_3^-]\ p}{1.2 \text{ mmol/liter}}$$

Therefore the arterial plasma bicarbonate concentration is normally about 24 mmol/liter because the logarithm of 20 is equal to 1.3.

Note that the term *total CO_2* refers to the dissolved carbon dioxide (including carbonic acid) *plus* the carbon dioxide present as bicarbonate.

A useful way to display the interrelationships between the variables of pH, P_{CO_2}, and bicarbonate concentration of the plasma, as expressed by the Henderson-Hasselbalch equation, is the *pH-bicarbonate diagram* shown in Fig. 8-1.

As can be seen from Fig. 8-1, pH is on the abscissa of the pH-bicarbonate diagram and the plasma bicarbonate concentration in millimoles per liter is on the ordinate. For each value of pH and bicarbonate ion concentration, there is a single corresponding P_{CO_2} on the graph. Conversely, for any particular pH and P_{CO_2}, only one bicarbonate ion concentration will satisfy the Henderson-Hasselbalch equation. If the P_{CO_2} is held constant, e.g., at 40 torr, an *isobar* line can be constructed, connecting the resulting points as the pH is varied. The representative isobars shown in Fig. 8-1 give an indication of the potential alterations of acid-base status when alveolar ventilation is increased or decreased. If everything else remains constant, hypoventilation leads to acidosis; hyperventilation leads to alkalosis.

The buffer value of the bicarbonate system *without the presence of hemoglobin* is about −5.4 mmol/liter/pH unit. That is, the bicarbonate concentration increases only 5.4 mmol/liter as enough acid in the form of carbon dioxide is added to the bicarbonate system to lower the pH by one unit. The rise in bicarbonate concentration represents the amount of carbonic acid added to the system. The bicarbonate buffer system is therefore a poor buffer for carbonic acid. The presence of hemoglobin makes blood a much better buffer, as can be seen in Fig. 8-2.

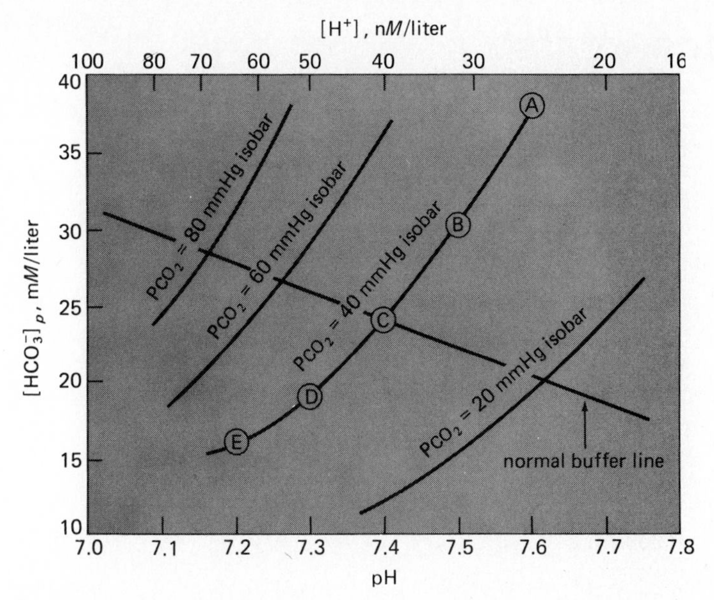

Figure 8-1 The pH-bicarbonate diagram with P_{CO_2} isobars. Note the hydrogen ion concentration in nanomoles per liter at the top of the figure corresponding to the pH's on the abscissa. Points A–E correspond to different pH's and bicarbonate concentrations all falling on the same P_{CO_2} isobar. (*Reprinted from Davenport, 1974, by permission of the University of Chicago Press.*)

The figure shows that increasing the hemoglobin concentration makes the buffering curve get steeper. That is, the bicarbonate concentration rises more at greater hemoglobin concentrations as carbonic acid (in the form of CO_2) is added to the blood. The increase in bicarbonate concentration is greater with more hemoglobin because as carbon dioxide is added to whole blood, the hydrogen ions formed by the dissociation of carbonic acid are buffered by hemoglobin (as will be discussed shortly). Most of the bicarbonate ions formed by this dissociation can therefore diffuse into the plasma. The buffer value of plasma in the presence of hemoglobin is therefore four to five times that of plasma separated from erythrocytes.

Phosphate

The phosphate buffer system mainly consists of the buffer pair of the dihydrogen phosphate ($H_2PO_4^-$) and the monohydrogen phosphate (HPO_4^{2-}) anions:

$$H_2PO_4^- \rightleftharpoons H^+ + HPO_4^{2-}$$

The pK of the acid form is 6.8, so that in pH's ranging near 7.0 the acid form can readily donate a proton and the base form can accept a proton.

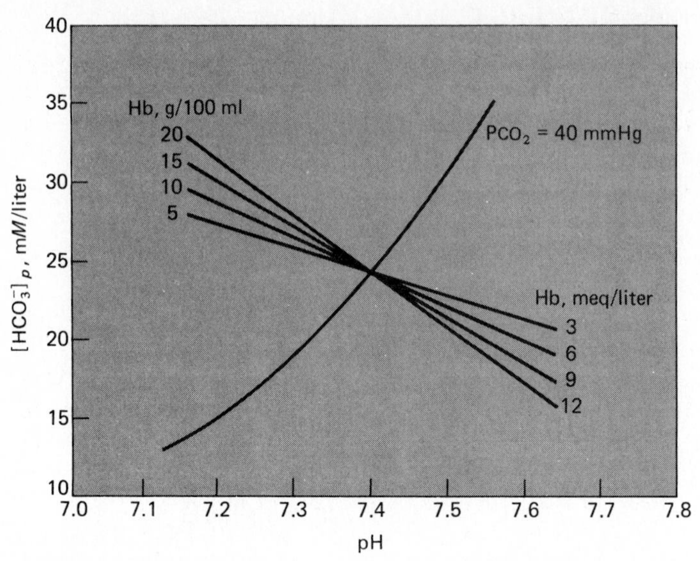

Figure 8-2 Buffer curves of plasma of blood containing 5, 10, 15, or 20 g of hemoglobin per 100 ml. (*Reprinted from Davenport, 1974, by permission of the University of Chicago Press.*)

Many organic phosphates found in the body also have p*K*'s within ±0.5 pH units of 7.0, and these compounds can also function as buffers under physiological conditions. These organic phosphates include such compounds as glucose-1-phosphate and adenosine triphosphate.

Proteins

Although several potential buffering groups are found on proteins, only one large group has p*K*'s in the pH range encountered in the blood. These are the imidazole groups in the histidine residues of the peptide chains. The p*K*'s of the various histidine residues on the different *plasma proteins* range from about 5.5 to about 8.5, thus providing a broad spectrum of buffer pairs. The protein present in the greatest quantity in the blood is *hemoglobin*. Thirty-six of the 540 amino acid residues in hemoglobin are histidine, with p*K*'s ranging from 7 to 8; the N-terminal valine residues also have a p*K* of about 7.8. As already noted, deoxyhemoglobin is a weaker acid than is oxyhemoglobin. That is, the p*K* of an imidazole group of one of the histidine residues in deoxygenated hemoglobin is higher than it is in the oxyhemoglobin state. Thus, as oxygen leaves hemoglobin in the tissue capillaries, the imidazole group removes hydrogen ions from the erythrocyte interior, allowing more carbon dioxide to be transported as bicarbonate. This process is reversed in the lungs.

Buffers of the Interstitial Fluid

The bicarbonate buffer system is the major buffer found in the interstitial fluid, including the lymph. The phosphate buffer pair is also found in the interstitial fluid. The volume of the interstitial compartment is much larger than that of the plasma, and so the interstitial fluid may play an important role in buffering.

Bone

The extracellular portion of bone contains very large deposits of calcium and phosphate salts, mainly in the form of hydroxyapatite. Although bone growth in a child causes a net production of hydrogen ions, in an otherwise healthy adult, where bone growth and resorption are in a steady state, bone salts can buffer hydrogen ions in chronic acidosis. Chronic buffering of hydrogen ions by the bone salts may therefore lead to demineralization of the bone.

Intracellular Buffering

The intracellular proteins and organic phosphates of most cells can function to buffer both fixed acids and carbonic acid. Again, this is largely a function of the histidine groups on the proteins and phosphate groups on such compounds as ATP and glucose-1-phosphate.

ACIDOSIS AND ALKALOSIS

Acid-base disorders can be divided into four major categories: respiratory acidosis, respiratory alkalosis, metabolic acidosis, and metabolic alkalosis. These primary acid-base disorders may occur singly ("simple") or in combination ("mixed") or may be altered by compensatory mechanisms.

Respiratory Acidosis

The arterial P_{CO_2} is normally kept at or near 40 torr (normal range is 35 to 45 torr by convention) by the mechanisms that regulate breathing. As will be discussed in Chap. 9, sensors exposed to the arterial blood and to the cerebrospinal fluid provide the central controllers of breathing with the information necessary to regulate the arterial P_{CO_2} at or near 40 torr. Any short-term alterations (i.e., those that occur without renal compensation, as will be discussed later in this chapter) in alveolar ventilation that result in an increase in alveolar and therefore arterial P_{CO_2} tend to lower the arterial pH, resulting in respiratory *acidosis*. This can be seen by looking at the $P_{CO_2} = 60$ torr and $P_{CO_2} = 80$ torr isobars in Fig. 8-1. The exact arterial pH at any Pa_{CO_2} depends on the bicarbonate and other buffers present in the blood. Pure changes in arterial P_{CO_2} caused by changes in ventilation travel along the normal in vivo buffer line, as shown in Figs. 8-1 and 8-3. This is similar to the in vitro plasma of blood buffer line at 15 g hemoglobin per 100 ml blood seen in Fig. 8-2. Pure uncompensated respiratory acidosis would correspond with point C on Fig. 8-3: at the intersection of an elevated P_{CO_2} isobar and the normal buffer line.

In respiratory acidosis the *ratio* of bicarbonate to CO_2 decreases. Yet, as can be seen at point C in Fig. 8-3, in uncompensated primary (simple) respiratory acidosis the absolute plasma bicarbonate concentration does increase somewhat because of the buffering of some of the hydrogen ions liberated by the dissociation of carbonic acid by nonbicarbonate buffers.

Any impairment of alveolar ventilation can cause respiratory acidosis. Depression of the respiratory centers in the medulla (see Chap. 9) by anesthetic agents, narcotics, hypoxia, central nervous system disease or trauma, or even greatly elevated Pa_{CO_2} *itself,* results in hypoventilation and respiratory acidosis. Interference with the neural transmission to the respiratory muscles by disease processes or by drugs or toxins or dysfunctions or deformities of the respiratory muscles or the chest wall can result in respiratory acidosis, as can restrictive, obstructive, and obliterative diseases of the lungs.

Respiratory Alkalosis

Alveolar ventilation in excess of that needed to keep pace with body carbon dioxide production results in alveolar and arterial P_{CO_2}'s below 35 torr. Such hyperventilation leads to respiratory alkalosis. As was the case

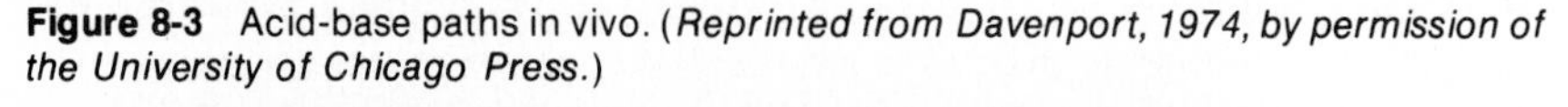

Figure 8-3 Acid-base paths in vivo. (*Reprinted from Davenport, 1974, by permission of the University of Chicago Press.*)

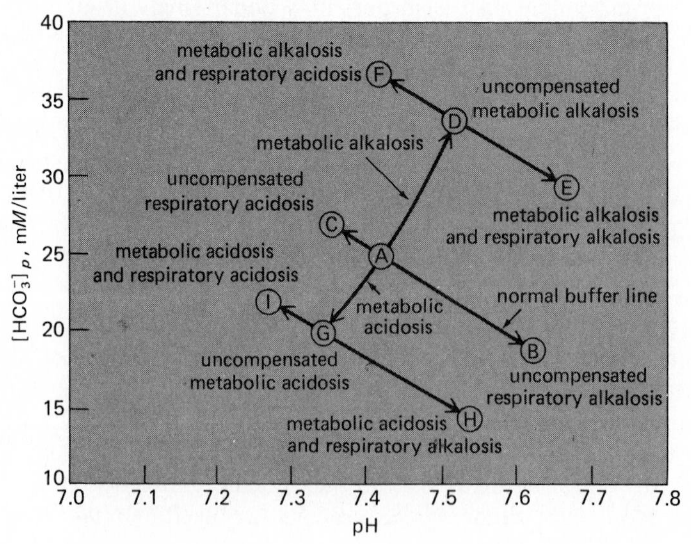

with primary respiratory acidosis, uncompensated primary respiratory alkalosis results in movement to a lower P_{CO_2} isobar along the normal buffer line, as seen at point B in Fig. 8-3. The decreased Pa_{CO_2} shifts the equilibrium of the series of reactions describing CO_2 hydration and carbonic acid dissociation to the left. This results in a decreased arterial hydrogen ion concentration, raising the pH, and a decreased plasma bicarbonate concentration. The ratio of bicarbonate to CO_2 increases.

The causes of respiratory alkalosis include anything leading to hyperventilation. *Hyperventilation syndrome,* a psychological dysfunction of unknown cause, results in chronic or recurrent episodes of hyperventilation and respiratory alkalosis. Drugs, hormones (such as progesterone), toxic substances, central nervous system diseases or disorders, bacteremias, fever, overventilation by mechanical ventilators (or the physician), or ascent to high altitude may all result in respiratory alkalosis.

Metabolic Acidosis

Metabolic acidosis may be more properly referred to as *nonrespiratory acidosis*. That is, it does not always involve aberrations in metabolism. Metabolic acidosis can be caused by the ingestion, infusion, or production of a fixed acid; by decreased renal excretion of hydrogen ions; by the movement of hydrogen ions from the intracellular to the extracellular compartment; or by the loss of bicarbonate or other bases from the extracellular compartment. As can be seen in Fig. 8-3, primary uncompensated metabolic acidosis results in a downward movement along the $P_{CO_2} = 40$ torr isobar to point G. That is, a net loss of buffer establishes a new blood buffer line lower than and parallel to the normal blood buffer line. P_{CO_2} is unchanged, hydrogen ion concentration is increased, and the ratio of bicarbonate concentration to CO_2 is decreased.

Ingestion of methyl alcohol or salicylates can cause metabolic acidosis by increasing the fixed acids in the blood. (Salicylate poisoning, e.g., aspirin overdose, causes both metabolic acidosis and later respiratory alkalosis.) Diarrhea can cause very great bicarbonate losses, resulting in metabolic acidosis. Renal dysfunctions can lead to an inability to excrete hydrogen ions, as well as an inability to reabsorb bicarbonate ions, as will be discussed in the next section. True "metabolic" acidosis may be caused by an accumulation of lactic acid in severe hypoxemia or shock and by diabetic ketoacidosis.

Metabolic Alkalosis

Metabolic, or nonrespiratory, alkalosis occurs when there is an excessive loss of fixed acids from the body or may occur as a consequence of the ingestion, infusion, or excessive renal reabsorption of bases such as bicarbonate. Figure 8-3 shows that primary uncompensated metabolic alkalosis results in an upward movement along the $P_{CO_2} = 40$ isobar to point D.

That is, a net gain of buffer establishes a new blood buffer line higher than and parallel to the normal blood buffer line. P_{CO_2} is unchanged, hydrogen ion concentration is decreased, and the ratio of bicarbonate concentration to CO_2 is increased.

Loss of gastric juice by vomiting results in a loss of hydrogen ions and may cause metabolic alkalosis. Excessive ingestion of bicarbonate or other bases (e.g., stomach antacids) or overinfusion of bicarbonate by the physician may cause metabolic alkalosis. In addition, diuretic therapy, treatment with steroids (or the overproduction of endogenous steroids), and conditions leading to severe potassium depletion may also cause metabolic alkalosis.

RESPIRATORY AND RENAL COMPENSATORY MECHANISMS

Uncompensated primary acid-base disturbances such as those indicated by points B, C, D, and G on Fig. 8-3 are seldom seen because respiratory and renal compensatory mechanisms are called into play to offset these disturbances. The two main compensatory mechanisms are functions of the respiratory and renal systems.

Respiratory Compensatory Mechanisms

The respiratory system can compensate for metabolic acidosis or alkalosis by altering alveolar ventilation. As discussed in Chap. 3, if carbon dioxide production is constant, the alveolar P_{CO_2} is roughly inversely proportional to the alveolar ventilation. This can be seen in the upper portion of Fig. 3-9. In metabolic acidosis, the elevated blood hydrogen ion concentration stimulates *chemoreceptors,* which, in turn, increase alveolar ventilation, thus decreasing arterial P_{CO_2}. This causes an increase in arterial pH, returning it toward the normal. (The mechanisms by which ventilation is regulated are discussed in detail in Chap. 9.) These events can be better understood by looking at Fig. 8-3. Point G represents uncompensated metabolic acidosis. As the respiratory compensation for the metabolic acidosis occurs, in the form of an increase in ventilation, the arterial P_{CO_2} falls. The point representing blood pHa, Pa_{CO_2}, and bicarbonate concentration would then move a short distance along the lower-than-normal buffer line (from point G toward point H) until a new lower Pa_{CO_2} is attained. This returns the arterial pH *toward* normal; complete compensation does not occur. Of course, the respiratory compensation for metabolic acidosis occurs almost simultaneously with the development of the acidosis. The blood pH, P_{CO_2}, and bicarbonate concentration point does not really move first from the normal (point A) to point G and *then* move a short distance along line GH; instead the compensation begins to occur as the acidosis develops so that the point takes an intermediate pathway between the two lines.

The respiratory compensation for metabolic alkalosis is to decrease alveolar ventilation, thus raising Pa_{CO_2}. This decreases arterial pH toward the normal, as can be seen on Fig. 8-3. Point D represents uncompensated metabolic alkalosis; respiratory compensation would move the blood pHa, Pa_{CO_2}, and bicarbonate concentration point a short distance along the new higher-than-normal blood buffer line toward point F. Again the compensation occurs as the alkalosis develops, with the point moving along an intermediate course.

Under most circumstances the *cause* of respiratory acidosis or alkalosis is a dysfunction in the ventilatory control mechanism or the breathing apparatus itself. Compensation for acidosis or alkalosis in these conditions must therefore come from outside the respiratory system. The respiratory compensatory mechanism can operate very rapidly (within minutes) to partially correct metabolic acidosis or alkalosis.

Renal Compensatory Mechanisms

The kidneys can compensate for respiratory acidosis and metabolic acidosis of nonrenal origin by excreting fixed acids and by retaining filtered bicarbonate. The kidneys can also compensate for respiratory alkalosis or metabolic alkalosis of nonrenal origin by decreasing hydrogen ion excretion and by decreasing the retention of filtered bicarbonate.

Renal Mechanisms in Acidosis The renal tubular cells secrete hydrogen ions into the tubular fluid. This is apparently accomplished by the generation of hydrogen ions and bicarbonate ions within the cell by the dissociation of carbonic acid. The carbonic acid is formed by the hydration of carbon dioxide via the carbonic anhydrase reaction, as shown in Fig. 8-4*A*. The carbon dioxide may be metabolically produced by the tubular cell itself or may be carried dissolved into the tubular fluid after production elsewhere in the body. The hydrogen ion generated by this process is transported into the tubular lumen and the bicarbonate ion is "reabsorbed" into the peritubular capillary. Sodium ions in the tubular fluid are exchanged for the hydrogen ions secreted into the tubular fluid to maintain electrical neutrality. There is also an interrelationship between renal potassium ion secretion and renal hydrogen ion secretion: When the secretion of one of these ions is elevated, secretion of the other is depressed. For this reason, disturbances in acid-base balance are usually associated with alterations in potassium ion balance and vice versa. The hydrogen ion secreted into the tubular lumen is buffered by tubular bicarbonate, phosphate, or the small quantities of other buffers found in the tubular fluid.

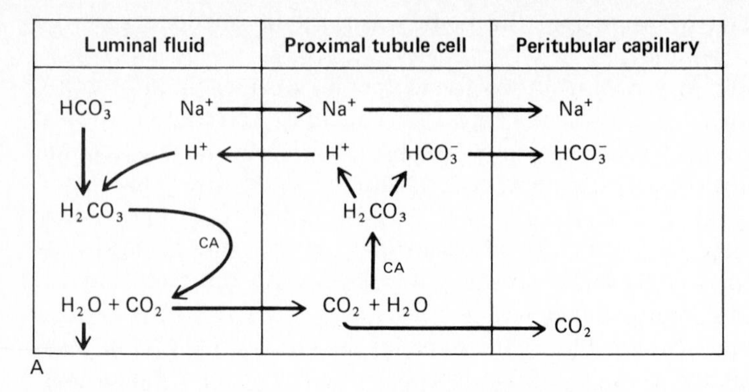

A

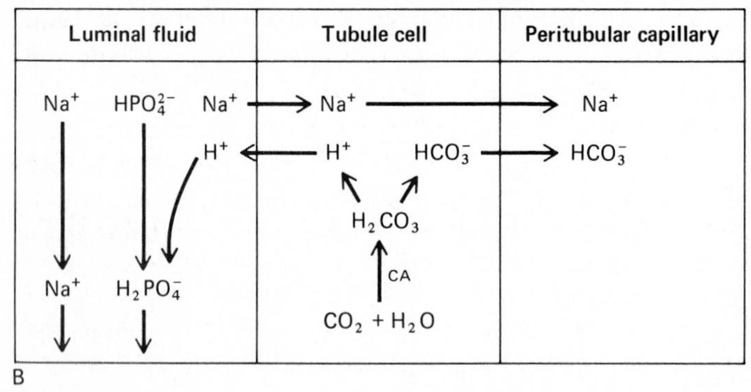

B

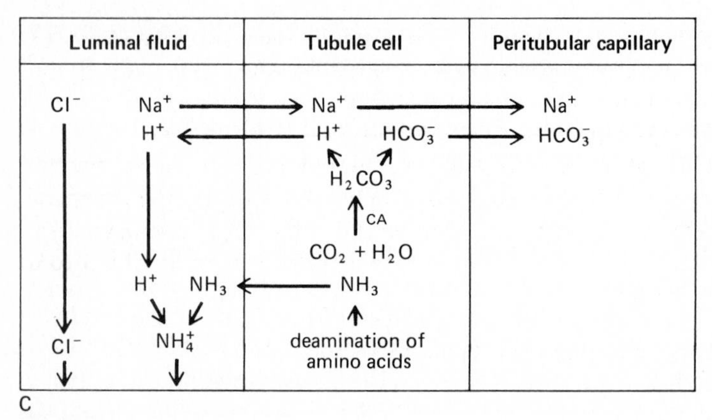

C

Figure 8-4 Schematic representation of renal fixed acid excretion and bicarbonate retention.

When a hydrogen ion combines with a tubular bicarbonate ion, forming carbonic acid, it may be converted to carbon dioxide by the carbonic anhydrase in the brush border of the proximal tubular cell. It is the *other* bicarbonate ion formed in the tubular cell that is reabsorbed into the peritubular capillary. Excretion of hydrogen ions and reabsorption of bicarbonate ions in this manner are schematized in Fig. 8-4*A*. About 90 percent of all filtered bicarbonate ions are reabsorbed in the proximal tubule either by *direct transport* from the tubular fluid or by this mechanism.

The remaining 10 percent of bicarbonate ions is either reabsorbed by this mechanism in the distal tubules and collecting duct or in the process of titration of tubular phosphate ions, as shown in Fig. 8-4*B*, or by the generation of ammonium ions, as shown in Fig. 8-4*C*. In each mechanism a bicarbonate ion is returned to the peritubular capillary.

Ammonia (NH_3) is actively formed in the renal cells by the deamination of amino acids such as glycine, alanine, aspartic acid, and leucine, as well as glutamine. Ammonia then diffuses into the tubular lumen and combines with a hydrogen ion, forming ammonium (NH_4^+) ions. Ammonium ions do not readily diffuse back into the renal tubular cells and they are highly soluble in tubular fluid; furthermore, ammonium is a very weak acid at tubular fluid pH's. The p*K* of the system is about 9.3.

Normally about 50 meq of hydrogen ion is secreted and about 50 meq of bicarbonate is reabsorbed daily by the kidneys. This process can increase during acidosis to the extent that the urine can be acidified to pH's as low as 4 to 5. This is about 800 times as acidic as normal plasma.

Renal Mechanisms in Alkalosis In alkalotic states the kidney decreases the secretion of hydrogen ions and decreases bicarbonate reabsorption. The kidney tends to reabsorb almost all of the filtered bicarbonate until the plasma bicarbonate concentration reaches about 27 to 28 meq/liter (normally it is about 24 meq/liter). Plasma bicarbonate is excreted above this threshold.

Time Course of Renal Mechanisms Renal compensatory mechanisms for acid-base disturbances operate much more slowly than respiratory compensatory mechanisms. For example, the renal compensatory responses to sustained respiratory acidosis or alkalosis may take 3 to 6 *days* to be completed.

Summary of Renal and Respiratory Contributions to Acid-Base Balance

The kidneys help to regulate acid-base balance by altering the excretion of fixed acids and the retention of the filtered bicarbonate; the respiratory system helps to regulate body acid-base balance by adjusting alveolar ventilation to alter alveolar P_{CO_2}. For these reasons some authors suggest that the Henderson-Hasselbalch equation is in effect

$$\text{pH} = \text{a constant} + \frac{\text{kidneys}}{\text{lungs}}$$

CLINICAL INTERPRETATION OF BLOOD GASES AND ACID-BASE STATUS

Samples of arterial blood are usually analyzed clinically to determine the "arterial blood gases": the arterial P_{O_2}, P_{CO_2}, and pH. The plasma bicarbonate can then be calculated from the pH and P_{CO_2} by using the Henderson-Hasselbalch equation. This can be done directly; or by using a nomogram or a special slide rule; or by graphical analysis such as the pH-bicarbonate diagram (the "Davenport plot," after its popularizer), the pH-P_{CO_2} diagram (the "Siggaard-Andersen"), or the composite acid-base diagram. Several programmable calculators have preprogrammed blood gas packages available, and many blood gas analyzers perform these calculations automatically.

Figure 8-5 shows the normal ranges of pHa, Pa_{CO_2}, and plasma bicarbonate concentration on a pH-bicarbonate diagram. Points that fall outside of these normal ranges indicate problems in acid-base balance. A thorough understanding of the patterns shown in Fig. 8-5 coupled with knowledge of a patient's P_{O_2} and other clinical findings can reveal a great deal about the underlying pathophysiological processes in progress. This can be seen in the sample problems for this chapter.

Base Excess

Calculation of the *base excess* or *base deficit* may be very useful in determining the therapeutic measures to be administered to a patient. The base

Figure 8-5 The pH-bicarbonate diagram with normal values indicated. Note that the normal range for arterial P_{CO_2} was considered to be 35 to 48 torr in the source from which this figure was reproduced. (*Reprinted from Davenport, 1974, by permission of the University of Chicago Press.*)

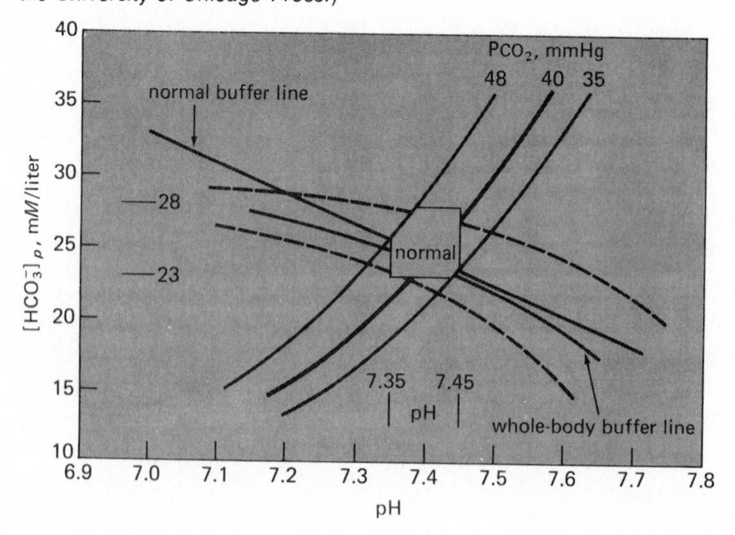

excess or base deficit is the number of milliequivalents of acid or base needed to titrate 1 liter of blood to pH 7.4 at 37° *if the* Pa_{CO_2} *were held constant at 40 torr.* It is not, therefore, just the difference between the plasma bicarbonate concentration of the sample in question and the normal plasma bicarbonate concentration because respiratory adjustments also cause a change in bicarbonate concentration: The arterial P_{CO_2} must be considered. Base excess can be determined by actually titrating a sample or by using a nomogram, diagram, special slide rule, or calculator program. The base excess is expressed in milliequivalents per liter above or below the normal buffer-base range—it therefore has a normal value of 0 ± 2 meq/liter. A base deficit is also called a *negative base excess.*

The base deficit can be used to estimate how much sodium bicarbonate (in milliequivalents) should be given to a patient by multiplying the base deficit (in milliequivalents per liter) times the patient's estimated extracellular fluid (ECF) space (in liters), which is the distribution space for the bicarbonate. The ECF is usually estimated to be 0.3 times the lean body mass in kilograms.

Chapter 9

The Control of Breathing

OBJECTIVES

The student understands the organization and function of the respiratory control system.

1. Describes the general organization of the respiratory control system.
2. Localizes the centers that generate the spontaneous rhythmicity of breathing.
3. Describes the groups of neurons that effect inspiration and expiration.
4. Describes the other centers in the brainstem that may influence the spontaneous rhythmicity of breathing.
5. Lists the cardiopulmonary and other reflexes that influence the breathing pattern.
6. States the ability of the brain cortex to temporarily override the normal pattern of inspiration and expiration.
7. Describes the effects of alterations in body oxygen, carbon dioxide, and hydrogen ion levels on the control of breathing.
8. Describes the sensors of the respiratory system for oxygen, carbon dioxide, and hydrogen ion concentration.

Breathing is spontaneously initiated in the central nervous system. A cycle of inspiration and expiration is *automatically* generated by neurons located in the brainstem; in eupneic states, breathing occurs without a conscious initiation of inspiration and expiration. Normal individuals do not have to worry about forgetting to breathe while they sleep.

This spontaneously generated cycle of inspiration and expiration can be modified, altered, or even temporarily suppressed by reflexes arising in the lungs and airways; or those arising in the cardiovascular system; or from information from receptors in contact with the cerebrospinal fluid; or by higher centers of the brain, such as the hypothalamus, the centers of speech, or other areas in the cortex. The centers that are responsible for the generation of the spontaneous rhythmicity of inspiration and expiration must therefore be able to alter their activity to meet the increased metabolic demand on the respiratory system during exercise or even be temporarily superseded or suppressed during speech or breath-holding.

The respiratory control centers in the brainstem effect the control of breathing via a "final common pathway" consisting of the spinal cord, the innervation of the muscles of respiration, such as the phrenic nerves, and the muscles of respiration themselves. Alveolar ventilation is therefore determined by the *interval* between successive groups of discharges of the respiratory neurons and the innervation of the muscles of respiration, which determines the respiratory rate or breathing frequency, and by the *frequency* of neural discharges transmitted by *individual* nerve fibers to their motor units, the *duration* of these discharges, and by the *number of motor units activated* during each inspiration or expiration, which determines the *depth* of respiration or the tidal volume.

Investigation of how the respiratory control system responds to stimuli such as exercise, hypoxia, hypercapnia, or drugs is relatively easy; investigation of how information from various sensors in the body and commands from higher brain centers are integrated with the activity of the respiratory neurons responsible for the spontaneous rhythmicity of breathing is much more difficult. This is because current research techniques are primarily limited to histological studies, to section and ablation experiments, and to recording or stimulating respiratory neuron activity. For these reasons, a discussion of the respiratory controller itself is mainly based on an inferential and deductive knowledge of the components of the system.

THE GENERATION OF SPONTANEOUS RHYTHMICITY

Studies done in the early part of the twentieth century demonstrated conclusively that the centers initiating breathing are located in the reticular formation of the medulla, beneath the floor of the fourth ventricle. If the brainstem of an anesthetized animal is sectioned *above* this area, as seen in the transection labeled III in Fig. 9-1, a pattern of inspiration and expiration is maintained even if all other afferents to this area, including the vagi, are also severed. If the brainstem is transected *below* this area, as seen in the transection labeled IV in Fig. 9-1, breathing ceases. This area, known as the *medullary center* (or medullary respiratory center), was originally believed to consist of two discrete groups of respiratory neurons, the *inspiratory neurons* and the *expiratory neurons*. It was thought that when the inspiratory neurons fired, their activity was transmitted to the muscles of inspiration, initiating inspiration; when expiratory neurons fired, their activity was transmitted to the muscles of expiration, initiating expiration. When the inspiratory neurons discharged, their activity was conducted to the expiratory neuron pool via collateral fibers, and the activity of the expiratory neurons was inhibited. Similarly, when the expiratory neurons discharged, their activity was conducted to the inspiratory neuron pool via collateral fibers, and the activity of the inspiratory neurons was inhibited. The *reciprocal inhibition* of these two opposing groups of neurons was believed to be the source of the spontaneous respiratory rhythmicity. More recent studies of the medullary center have not entirely supported this early hypothesis. The medullary center is anatomically diffuse, but functionally integrated; the current understanding of its organization is discussed in the next section.

Figure 9-1 The effects of transections at different levels of the brainstem on the ventilatory pattern of anesthetized animals. *At left* is a schematic representation of the dorsal surface of the lower brainstem. *At right* is a schematic representation of the breathing patterns (inspiration is upward) corresponding to the transections with the vagus nerves intact or transected. DRG is the dorsal respiratory group; VRG is the ventral respiratory group; APC is the apneustic center; PNC is the pneumotaxis center; IC is the inferior colliculus; CP is the cerebral peduncle. [*From Berger, 1977. Reprinted by permission of the N. Engl. J. Med.*].

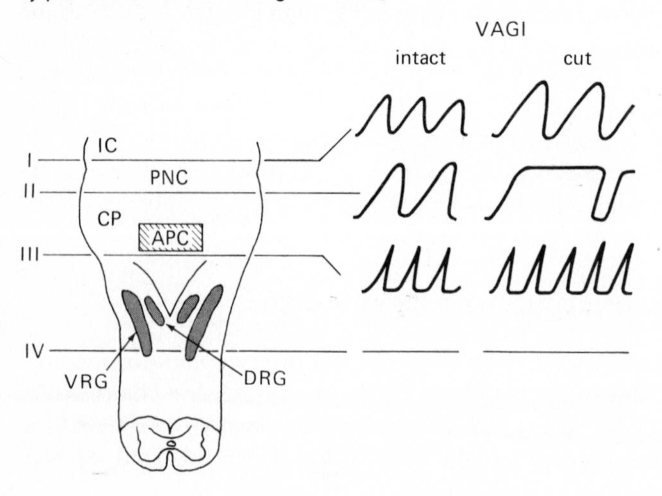

THE MEDULLARY RESPIRATORY CENTER

Recent studies have shown that the medullary respiratory center does not consist of a discrete "inspiratory center" and a discrete "expiratory center." There are two dense bilateral aggregations of respiratory neurons known as the *dorsal respiratory groups* (DRG in Figs. 9-1 and 9-2) and the *ventral respiratory groups* (VRG in Figs. 9-1 and 9-2). Inspiratory and expiratory neurons are anatomically intermingled to a greater or lesser extent within these areas.

The Dorsal Respiratory Group

The dorsal respiratory groups are located bilaterally in the *nucleus of the tractus solitarius* (NTS), as shown in Figs. 9-1 and 9-2. They consist mainly of *inspiratory cells*. These inspiratory neurons project primarily to the contralateral spinal cord. They probably serve as the principal initiators of the activity of the phrenic nerves and are therefore responsible for maintaining the activity of the diaphragm. Dorsal respiratory group neurons send collateral fibers to those in the ventral respiratory group, but the ventral respiratory group does not send collateral fibers to the dorsal respiratory group. Reciprocal inhibition therefore seems an unlikely explanation of the spontaneous inspiratory and expiratory rhythmicity. The nucleus of the tractus solitarius is the primary *projection site* of visceral afferent fibers of the ninth cranial nerve (the glossopharyngeal) and the

Figure 9-2 Schematic representation of the medullary respiratory groups. DRG is the dorsal respiratory group; VRG is the ventral respiratory group; NTS is the nucleus of the tractus solitarius; NA is the nucleus ambiguus; NRA is the nucleus retroambigualis; C1 is the first cervical root. See text for details. [*From Berger, 1977. Reprinted by permission of the N. Engl. J. Med.*].

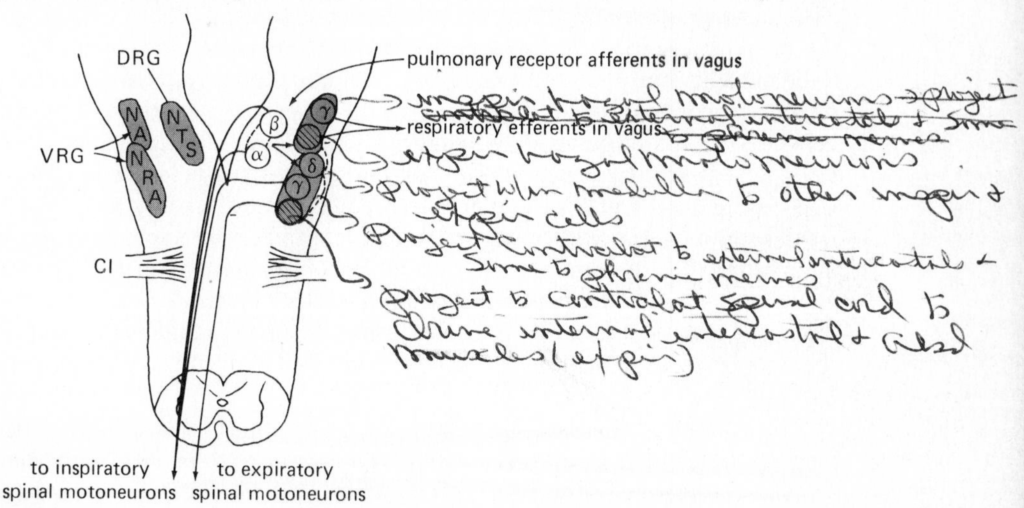

tenth cranial nerve (the vagus). These nerves carry information about the arterial P_{O_2}, P_{CO_2}, and pH from the carotid and aortic arterial chemoreceptors and information concerning the systemic arterial blood pressure from the carotid and aortic baroreceptors. In addition, the vagus carries information from stretch receptors and other sensors in the lungs that may also exert profound influences on the control of breathing. The effects of these chemoreceptors, baroreceptors, and other sensors on the control of breathing will be discussed in detail later in this chapter. The location of the dorsal respiratory group within the nucleus of the tractus solitarius suggests that the dorsal respiratory group may be the site of integration of various inputs that can reflexly alter the spontaneous pattern of inspiration and expiration.

There are two populations of inspiratory neurons in the dorsal respiratory group, as shown in Fig. 9-2. One population, called the *I alpha cells,* is inhibited by lung inflation; the second population, the *I beta cells,* is excited by lung inflation. These cells may play an important role in the Hering-Breuer reflexes described later in this chapter.

In summary, the dorsal respiratory group lies close to or is identical with the primary site of respiratory rhythm generation. It is probably responsible for driving the diaphragm and is probably also the origin of the rhythmic respiratory drive to the ventral respiratory group and many spinal motoneurons. In addition, the dorsal respiratory group is probably the initial integrating site for many cardiopulmonary reflexes that affect the respiratory rhythm.

The Ventral Respiratory Group

The ventral respiratory groups are located bilaterally in the *nucleus ambiguus* and the *nucleus retroambigualis,* as shown in Figs. 9-1 and 9-2. They consist of both inspiratory and expiratory neurons. Those neurons in the nucleus ambiguus are primarily vagal motoneurons, both inspiratory (labelled γ in Fig. 9-2) and expiratory (shaded). In the nucleus retroambigualis, the inspiratory cells are located more rostrally and the expiratory cells are located more caudally (shaded). There appear to be two populations of inspiratory cells in the nucleus retroambigualis: One group, labeled the γ cells in Fig. 9-2, mainly projects contralaterally to external intercostal muscles, with some fibers also sent to the phrenic nerves, thus innervating the diaphragm; the second group, labeled δ in the figure, appears to project only within the medulla to other inspiratory and expiratory cells. The expiratory neurons in the nucleus retroambigualis project to the contralateral spinal cord to drive the internal intercostal and abdominal muscles.

In summary, the ventral respiratory group neurons consist of both inspiratory and expiratory cells. Their major function is to drive either spinal respiratory neurons, innervating mainly the intercostal and abdominal muscles, or the auxiliary muscles of respiration innervated by the vagus nerves. The dorsal respiratory group sends fibers to the ventral respiratory group, but not vice versa. Therefore the dorsal respiratory group may drive the ventral respiratory group, but reciprocal inhibition between the two groups is unlikely. The ventral respiratory group does not appear to be an initial processing site for sensory information.

Because no reciprocal inhibitory patterns have been found between the dorsal and ventral respiratory groups, the older hypotheses of the generation of the spontaneous rhythmicity of inspiration and expiration do not appear to be valid. Many expiratory cells may not fire at all during the passive expirations seen in eupneic breathing (see Chap. 2); those that do discharge do not cause contraction of the expiratory muscles. It now appears that cells in or near to the dorsal respiratory group probably act as pacemakers themselves, with their activity modulated by afferent information and higher brain centers.

THE APNEUSTIC CENTER

If the brainstem is transected in the pons at the level denoted by the line labeled II in Fig. 9-1, a breathing pattern called *apneusis* results if the vagus nerves have also been transected. Apneustic breathing consists of prolonged inspiratory efforts interrupted by occasional expirations. Afferent information that reaches this *"apneustic center"* via the vagus nerves must be important in preventing apneusis because apneusis does not occur if the vagus nerves are intact, as shown schematically in Fig. 9-1.

Studies have indicated that apneusis is probably a general phenomenon involving a sustained discharge of medullary inspiratory neurons. Investigators believe that the apneustic center may be the site of the normal "inspiratory cutoff switch"; that is, it is the site of projection of various types of afferent information that can *terminate inspiration*. Apneusis is a result of the inactivation of the inspiratory cutoff mechanism. The specific group of neurons that function as the apneustic center have not been identified.

THE PNEUMOTAXIC CENTER

If the brainstem is transected immediately caudal to the inferior colliculus, as denoted by the line labeled I in Fig. 9-1, the breathing pattern shows an essentially normal balance between inspiration and expiration, even if the vagus nerves are transected. As discussed in the previous section, transections made caudal to the line labeled II in Fig. 9-1 lead to apneusis, in the absence of the vagus nerves. A group of respiratory neurons known as the *"pneumotaxic center"* therefore functions to modulate the activity of the apneustic center. These cells, located in the upper pons in the *nucleus parabrachialis medialis* and the *Kolliker-Fuse nucleus,* probably function to "fine-tune" the breathing pattern. Electrical stimulation of these structures can result in synchronization of phrenic nerve activity with the stimulus or premature switching from inspiration to expiration and vice versa. Pulmonary inflation afferent information can inhibit the activity of the pneumotaxic center, which may in turn act to modulate the threshold for lung inflation inspiratory cutoff. The pneumotaxic center may also modulate the respiratory control system's response to other stimuli such as hypercapnia or hypoxia.

SPINAL PATHWAYS

Axons projecting from the dorsal respiratory group, the ventral respiratory group, the cortex, and other supraspinal sites descend in the spinal white matter to influence the phrenics, and intercostal and abdominal muscles of respiration, as already discussed. At the level of these spinal respiratory motoneurons, there is integration of descending influences as well as the presence of local spinal reflexes that can affect these motoneurons. Descending axons with inspiratory activity excite external intercostal motoneurons and also inhibit internal intercostal motoneurons by exciting spinal inhibitory interneurons.

Ascending pathways in the spinal cord, carrying information from pain, touch, and temperature receptors, as well as from proprioceptors, can also influence breathing, as will be discussed in the next section. Inspiratory and expiratory fibers appear to be separated in the spinal cord.

REFLEX MECHANISMS OF RESPIRATORY CONTROL

A large number of sensors located in the lungs, the cardiovascular system, the muscles and tendons, and the skin and viscera can elicit reflex alterations in the control of breathing.

Respiratory Reflexes Arising From Pulmonary Stretch Receptors

Three respiratory reflexes can be elicited by stimulation of the pulmonary stretch receptors: the Hering-Breuer inflation reflex, the Hering-Breuer deflation reflex, and the "paradoxical" reflex of Head.

The Hering-Breuer Inflation Reflex In 1868 Breuer and Hering reported that a maintained distention of the lungs of anesthetized animals decreased the frequency of the inspiratory effort or caused a transient apnea. The stimulus for this reflex is pulmonary inflation. That is, the sensors are *stretch receptors* located within the smooth muscle of large and small airways. The afferent pathway consists of large myelinated fibers in the vagus; as mentioned previously, these fibers appear to enter the brainstem and project to the nucleus of the tractus solitarius and the pneumotaxic center. The efferent limb of the reflex consists of bronchodilation in addition to the apnea or slowing of the ventilatory frequency (due to an increase in the time spent in expiration) already mentioned. Lung inflation also causes reflex effects in the cardiovascular system: Moderate lung inflations cause an increase in heart rate and may cause a slight vasoconstriction; very large inflations may cause a decrease in heart rate and systemic vascular resistance.

The Hering-Breuer inflation reflex was originally believed to be an important determinant of the rate and depth of ventilation because of the original studies performed on anesthetized animals. Vagotomized anesthetized animals breathe much more deeply and less frequently than they did before their vagus nerves were transected. It was therefore assumed that the Hering-Breuer inflation reflex acts tonically to limit the tidal volume and establish the depth and rate of breathing. More recent studies on unanesthetized humans have cast doubt on this conclusion because the central threshold of the reflex is much higher than the normal tidal volume during eupneic breathing. Tidal volumes of 800 to 1000 ml are generally required to elicit this reflex in conscious eupneic men. The Hering-Breuer inflation reflex may help to minimize the work of breathing by inhibiting large tidal volumes (see Chap. 2) as well as prevent overdistention of the alveoli at high lung volumes. It may also be important in the control of breathing in babies.

The Hering-Breuer Deflation Reflex Breuer and Hering also noted that abrupt *deflation* of the lungs augments the ventilatory rate. This could be a result of *decreased* stretch receptor activity, the stimulation of as yet unknown receptors in the lungs, or the stimulation of other pulmonary receptors such as the irritant receptors and "J" receptors, which will be discussed later in this chapter. The afferent pathway is the vagus and the effect is hyperpnea. The importance of this reflex in human beings has not yet been demonstrated conclusively. It may be responsible for the increased ventilation elicited when the lungs are deflated abnormally, as in pneumothorax, or it may play a role in the periodic spontaneous deep breaths ("sighs") that help to prevent atelectasis. These sighs occur occasionally and irregularly during the course of normal, quiet, spontaneous breathing. They consist of a slow deep inspiration (larger than a normal tidal volume) followed by a slow deep expiration. This response appears to be very important because patients maintained on mechanical ventilators must be given periodic deep breaths or they develop diffuse atelectasis, which may lead to arterial hypoxemia.

The Paradoxical Reflex of Head In 1889 Henry Head performed experiments designed to show the effects of the Hering-Breuer inflation reflex on the control of breathing. Instead of transecting the vagus nerves, he decided to block their function by cooling them to 0°C. As he rewarmed the vagus nerves, he noted that in the situation of a selective *partial* block of the vagus nerves lung inflation caused a further inspiration instead of the apnea expected when the vagus nerves were completely functional. The receptors for this "paradoxical" reflex are located in the lungs, but their precise location is not known. Afferent information travels in the vagus; the effect is very deep inspirations. This reflex may also be involved in the sigh response. It has also been suggested that this reflex may be involved in generating the first breath of the newborn baby; very great inspiratory efforts must be generated to inflate the collapsed lungs.

Respiratory Reflexes Arising from Irritant Receptors in the Airways and Lungs

Mechanical or chemical irritation of the airways (and possibly the alveoli) can elicit a reflex cough or sneeze or hyperpnea, bronchoconstriction, and increased blood pressure. The receptors are located in the nasal mucosa, upper airways, the tracheobronchial tree, and possibly in the alveoli themselves. The afferent pathways are the vagus nerves for all but the receptors located in the nasal mucosa, which send information centrally via the trigeminal and olfactory tracts. The cough and the sneeze reflexes are discussed in greater detail in Chap. 10.

Respiratory Reflexes Arising from Pulmonary Vascular Receptors (Type-J Receptors)

Pulmonary embolism causes apnea or rapid shallow breathing; pulmonary vascular congestion causes hyperpnea. Injection of chemicals such as phenyldiguanide or capsaicin into the pulmonary circulation may also elicit apnea or rapid shallow breathing. The receptors responsible for initiating these responses are believed to be located in the walls of the pulmonary capillaries, and hence they are called *type-J* (for juxtapulmonary-capillary) *receptors*. They are believed to be stimulated by pulmonary vascular congestion or an increase in pulmonary interstitial fluid volume and would therefore be responsible for the reflexes discussed above. In addition, they might also be responsible for the *dyspnea* (a feeling of difficult or labored breathing) encountered during the pulmonary vascular congestion and edema secondary to left ventricular failure or even the dyspnea that healthy people feel during exercise. The afferent pathway of these reflexes is slow-conducting nonmyelinated vagal fibers.

Respiratory Reflexes Arising from the Cardiovascular System

The arterial chemoreceptors, and to a much lesser extent the arterial baroreceptors, can exert great influences on the respiratory control system. The role of the arterial chemoreceptors in the control of ventilation will be discussed in great detail in subsequent sections of this chapter and will only be briefly summarized here.

Arterial Chemoreceptors The arterial chemoreceptors are located bilaterally in the carotid bodies, which are situated near the bifurcations of the common carotid arteries, and in the aortic bodies, which are located in the arch of the aorta. They respond to low arterial P_{O_2}'s, high arterial P_{CO_2}'s, and low arterial pH's, with the carotid bodies generally capable of a greater response than the aortic bodies. The afferent pathway from the carotid body is Hering's nerve, a branch of the glossopharyngeal nerve; the afferent pathway from the aortic body is the vagus. The reflex effects of stimulation of the arterial chemoreceptors are hyperpnea, bronchoconstriction, and increased blood pressure. The *direct* effect of arterial chemoreceptor stimulation is a decrease in heart rate; however, this is usually masked by an increase in heart rate secondary to the increase in lung inflation.

Arterial Baroreceptors The arterial baroreceptors exert a minor influence on the control of ventilation. Baroreceptors are stretch receptors that are responsive to changes in pressure. They are located in the carotid sinuses, which are situated at the origin of the internal carotid arteries near the bifurcation of the common carotid arteries, and in the aortic arch. The afferent pathways are Hering's nerve and the glossopharyngeal for the carotid baroreceptors, and the vagus nerve for the aortic baroreceptors. The effects of stimulation of the arterial baroreceptors by elevated blood pressure are apnea and bronchodilation.

Respiratory Reflexes Arising from Muscles and Tendons

Stimulation of receptors located in the muscles, tendons, and joints can increase ventilation. Included are receptors in the muscles of respiration (e.g., muscle spindles) and rib cage as well as other skeletal muscles, joints, and tendons. These receptors may play an important role in adjusting the ventilatory effort to elevated work loads and may help to minimize the work of breathing. They may also participate in initiating and maintaining the elevated ventilation that occurs during exercise, as will be discussed later in this chapter. Afferent information ascends to the respiratory controller via the spinal cord, as mentioned previously in this chapter.

Reflex Respiratory Responses to Pain

Somatic pain generally causes hyperpnea; visceral pain generally causes apnea or decreased ventilation. Physicians often make use of the ventilatory response to somatic pain to increase ventilation in patients with depressed breathing by slapping or pinching them.

INFLUENCES OF HIGHER CENTERS

The spontaneous rhythmicity generated in the medullary respiratory center can be completely overwhelmed (at least temporarily) by influences from higher brain centers. In fact, the greatest minute ventilations obtainable from healthy conscious human subjects can be attained *voluntarily,* exceeding those obtained with the stimuli of severe exercise, hypercapnia, or hypoxia. This is the underlying concept of the maximum voluntary ventilation (MVV) test often used to assess the respiratory system. Conversely, the respiratory rhythm can be *completely* suppressed for *several minutes* by voluntary breath-holding, until the chemical drive to respiration (high P_{CO_2} and low P_{O_2} and pH) overwhelms the voluntary suppression of breathing at the "breakpoint."

During speech, singing, or playing a wind instrument the normal cycle of inspiration and expiration is automatically modified by higher brain centers. In certain emotional states chronic hyperventilation severe enough to cause respiratory alkalosis may occur, as was discussed in Chap. 8.

THE RESPONSE TO CARBON DIOXIDE

The respiratory control system normally reacts very effectively to alterations in the internal "chemical" environment of the body. Changes in the body P_{CO_2}, pH, and P_{O_2} result in alterations in alveolar ventilation designed to return these variables to their normal values. Special neuronal units called *chemoreceptors* alter their activity when their own local chemical environment changes and can therefore supply the central respiratory controller with the afferent information necessary to make the proper adjustments in alveolar ventilation and change the whole-body P_{CO_2}, pH, and P_{O_2}. The respiratory control system, therefore, functions as a *negative-feedback system.*

The arterial and cerebrospinal fluid partial pressures of carbon dioxide are probably the most important inputs to the ventilatory control system in establishing the breath-to-breath levels of tidal volume and ventilatory frequency. (Of course, changes in carbon dioxide lead to changes in hydrogen ion concentration, and so these two stimuli can be difficult to separate.) An elevated level of carbon dioxide is a very powerful stimulus to ventilation: Only voluntary hyperventilation and the hyperpnea of exercise can surpass the minute ventilations obtained with hypercapnia. Yet the arterial P_{CO_2} is so precisely controlled that it changes little (<1 torr) during exercise severe enough to increase metabolic carbon dioxide production 10-fold.

The ventilatory responses of humans to elevated levels of carbon dioxide in the inspired air ($F_{I_{CO_2}}$) are shown in Fig. 9-3. The figure shows that elevating $F_{I_{CO_2}}$ increases the minute ventilation. The effect is most pronounced in $F_{I_{CO_2}}$'s in the range of 0.05 to 0.10 (5 to 10% CO_2 in inspired gas), which corresponds to alveolar P_{CO_2}'s between about 40 and 70 torr. Above 10 to 15% carbon dioxide in inspired air there is little further increase in alveolar ventilation: Very high arterial P_{CO_2}'s (>70 to 80 torr) may directly produce respiratory depression.

The physiological response to elevated carbon dioxide is dependent on its concentration. Low concentrations of carbon dioxide in the inspired air are easily tolerated, with an increase in ventilation the main effect. Above approximately 7% carbon dioxide in the inspired air, subjects report dyspnea, severe headaches secondary to the cerebral vasodilation caused by the elevated Pa_{CO_2}, restlessness, faintness, and dulling of consciousness. These reactions are in addition to the greatly elevated alveolar ventilation seen in Fig. 9-3. A loss of consciousness occurs at inspired CO_2 concentrations greater than 15%; muscular rigidity and tremors also occur at this level. With 20 to 30% inspired carbon dioxide generalized convulsions are produced almost immediately.

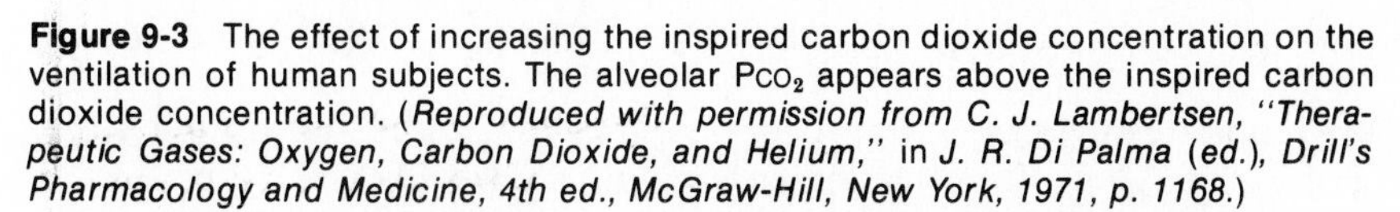

The ventilatory response of a normal conscious person to more phys-

Figure 9-3 The effect of increasing the inspired carbon dioxide concentration on the ventilation of human subjects. The alveolar P_{CO_2} appears above the inspired carbon dioxide concentration. (*Reproduced with permission from C. J. Lambertsen, "Therapeutic Gases: Oxygen, Carbon Dioxide, and Helium," in J. R. Di Palma (ed.), Drill's Pharmacology and Medicine, 4th ed., McGraw-Hill, New York, 1971, p. 1168.*)

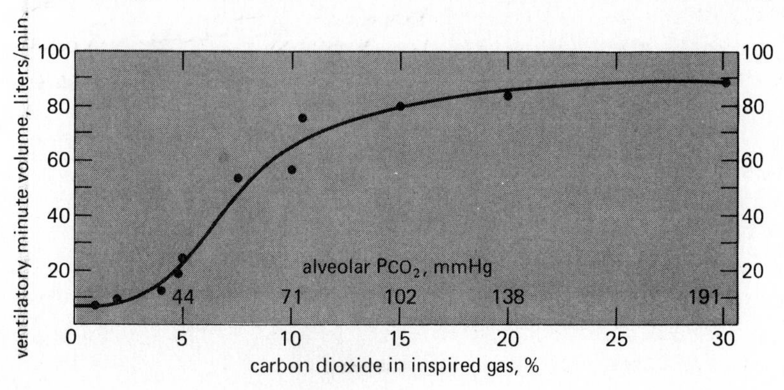

iological levels of carbon dioxide is shown schematically in Fig. 9-4. Inspired concentrations of carbon dioxide or metabolically produced carbon dioxide producing alveolar (and arterial) P_{CO_2}'s in the range of 38 to 50 torr linearly increase alveolar ventilation. The slope ($\Delta\dot{V}_E/\Delta P_{CO_2}$) is quite steep; it varies from person to person, with a slope of 2 to 5 liters/min per torr of $P_{A_{CO_2}}$ (the mean is equal to 2.0 to 2.5 liters/min per torr $P_{A_{CO_2}}$) for young healthy adults. The slope appears to decrease with age.

Figure 9-4 also shows that *hypoxia* potentiates the ventilatory response to carbon dioxide. At lower arterial P_{O_2}'s (e.g., 35 and 50 torr) the response curve is shifted to the left and the slope is steeper. That is, for any particular arterial P_{CO_2}, the ventilatory response is greater at a lower arterial P_{O_2}. This may be caused by the effects of hypoxia at the chemoreceptor itself; changes in the central acid-base status secondary to hypoxia may also contribute to the enhanced response.

Other influences on the carbon dioxide response curve are shown schematically in Fig. 9-5. Sleep shifts the curve slightly to the right. The arterial P_{CO_2} normally increases during slow wave sleep, rising as much as 5 to 6 torr during deep sleep. A depressed response to carbon dioxide during sleep may be involved in *sleep apnea,* a condition characterized by abnormally long periods (1 to 2 min) between breaths during sleep. This is a potentially dangerous condition in both infants and adults. Narcotics and anesthetics may profoundly depress the ventilatory response to carbon dioxide. Indeed, respiratory depression is the most common cause of death in cases of overdose of opiate alkaloids and their derivatives, barbiturates, and most anesthetics. Endorphins probably also depress the response to carbon dioxide. Chronic obstructive lung diseases also de-

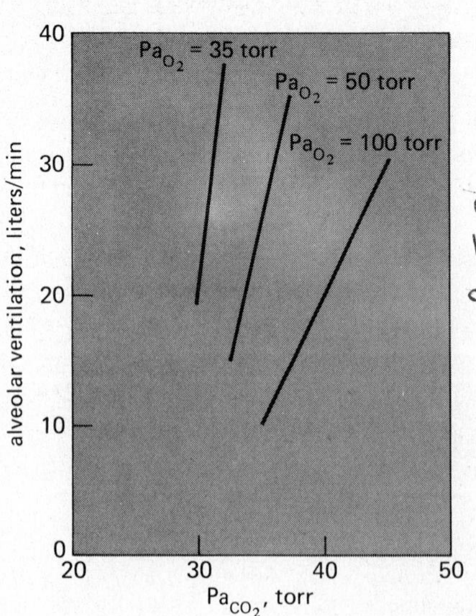

Figure 9-4 Schematic representation of ventilatory carbon dioxide response curves at three different levels of arterial P_{O_2}.

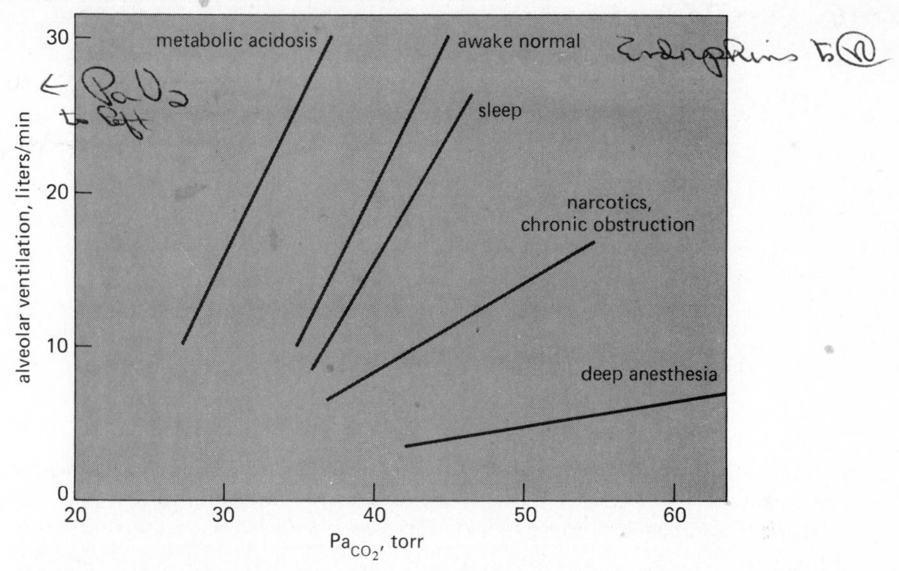

Figure 9-5 Schematic representation of the effects of sleep, narcotics, chronic pulmonary obstructive disease, deep anesthesia, and metabolic acidosis on the ventilatory response to carbon dioxide.

press the ventilatory response to hypercapnia. This reaction, however, may only be in part caused by depressed ventilatory drive secondary to central acid-base changes. Ventilatory drive may be elevated with increased Pa_{CO_2}, but the work of breathing may be so great that ventilation cannot increase. Metabolic acidosis displaces the carbon dioxide response curve to the left, indicating that for any particular Pa_{CO_2} ventilation is increased during metabolic acidosis.

As already discussed, the respiratory control system constitutes a negative-feedback system. This is exemplified by the response to carbon dioxide. Increased *metabolic production* of carbon dioxide increases the carbon dioxide brought to the lung. If alveolar ventilation stayed constant, the alveolar P_{CO_2} would increase, as would arterial and cerebrospinal P_{CO_2}. This stimulates alveolar ventilation. Increased alveolar ventilation decreases alveolar and arterial P_{CO_2}, as was discussed in Chap. 3 (see Fig. 3-9), returning the P_{CO_2} to the original value, as shown in Fig. 9-6.

The curve labeled A in Fig. 9-6 shows the effect of increasing ventilation (here $\dot{V}_I$, or the inspired minute volume in liters per minute) on the arterial P_{CO_2}. Note that the independent variable for curve A is on the *ordinate* and the dependent variable is on the abscissa. This graph is really the same as that shown in the upper part of Fig. 3-9. Curve B is the steady-state ventilatory response to elevated arterial P_{CO_2}'s as obtained by increasing the percent inspired carbon dioxide—that is, it is a typical CO_2 response curve (like that seen in Fig. 9-4). The point at which the two

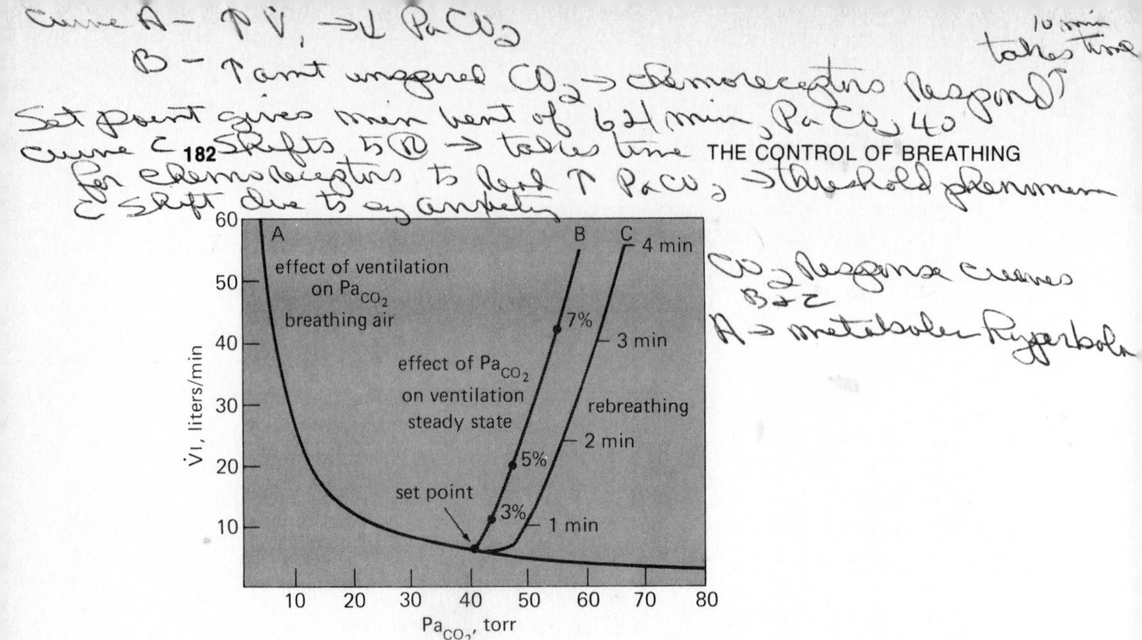

Figure 9-6 *Curve A.* The effect of ventilation on the arterial P_{CO_2}. Note that ventilation is the independent variable and PA_{CO_2} is the dependent variable. *Curve B.* The steady state ventilatory response to elevated PA_{CO_2} as obtained by breathing carbon dioxide mixtures. *Curve C.* The ventilatory response to rebreathing the gas from a 5- to 7-liter bag filled with 7% CO_2 and 40 to 93% O_2 for 4 min. See text for details. [*From Berger, 1977. Reprinted by permission of the N. Engl. J. Med.*].

curves cross is the "set point" for the system, normally a Pa_{CO_2} of 40 torr. Curve C shows the result of a "rebreathing" test in which a subject breathes for about 4 min into and out of a 5-liter gas bag prefilled with about 7% CO_2 and a high percentage of oxygen (40 to 93%). After a few seconds of rebreathing, the slope of curve C is parallel to curve B. Here the increasing Pa_{CO_2} is due to metabolically produced carbon dioxide. The rebreathing test generates a CO_2 response curve much faster than the steady-state method shown in Curve B.

As already mentioned, the respiratory control system constitutes a negative-feedback system with the P_{CO_2}, pH, and P_{O_2} the controlled variables. To act as a negative-feedback system, the respiratory controller must receive information concerning the levels of the controlled variables from sensors in the system. These sensors, or chemoreceptors, are located within the systemic arterial system and within the brain itself. The arterial chemoreceptors, which are often referred to as the *peripheral chemoreceptors,* are located in the carotid and aortic bodies; the *central chemoreceptors* are probably located bilaterally near the ventrolateral surface of the medulla in the brainstem. The peripheral chemoreceptors are exposed to arterial blood; the central chemoreceptors are exposed to cerebrospinal fluid. The central chemoreceptors are therefore on the *brain* side of the blood-brain barrier. Both the peripheral and central chemoreceptors respond to elevations in the partial pressure of carbon dioxide, although the response may possibly be related to the local increase in hydrogen ion concentration that occurs with elevated P_{CO_2}. That is, the sensors may be responding to the increased carbon dioxide concentration, the subsequent increase in hydrogen ion concentration, or both.

Peripheral Chemoreceptors

The peripheral chemoreceptors alter their firing rate in response to changes in the gas composition and hydrogen ion concentration of the arterial blood. Increased arterial P_{CO_2} or decreased arterial pH and P_{O_2} stimulates increased peripheral chemoreceptor activity. There is considerable impulse traffic in the afferent fibers from the arterial chemoreceptors at normal levels of arterial P_{O_2}, P_{CO_2}, and pH. The response of the receptors is both rapid enough and sensitive enough that they can relay information concerning breath-to-breath alterations in the composition of the arterial blood to the medullary respiratory center. Recordings made of afferent fiber activity have demonstrated increased impulse traffic in a single fiber to increased P_{CO_2} and decreased pH and P_{O_2}, although the sensors themselves may not react to all three stimuli. The afferent fibers all appear to have similar response curves to the three stimuli: Recruitment of new fibers does not appear to play a role in the response to the stimuli. The carotid bodies appear to exert a much greater influence on the respiratory controller than do the aortic bodies, especially with respect to decreased P_{O_2} and pH; the aortic bodies may exert a greater influence on the cardiovascular system.

The response of the arterial chemoreceptors changes *nearly linearly* with the arterial P_{CO_2} over the range of 20 to 60 torr. Hyperthermia also increases the activity of the peripheral chemoreceptors, as do hypotension and increased sympathetic stimulation of the vascular smooth muscle. The latter two conditions appear to act by reducing blood flow to the chemosensitive elements.

The exact mechanism by which the chemoreceptors operate is uncertain. The ultrastructure of the chemoreceptors is complicated, consisting of several types of so-called glomus cells with many nerve fibers in close proximity. It is not entirely clear whether it is the glomus cells or the fiber endings themselves that act as the chemosensitive elements. The glomus cells contain large amounts of catecholamines, especially dopamine. The glomus cells form synapses with the sensory nerve endings and possibly with each other. One current theory is that it is the nerve fibers that act as the chemoreceptors and the glomus cells are dopaminergic inhibitory interneurons.

The carotid body has a very great blood flow (estimated to be as great as 2000 ml per 100 g of tissue per minute) and a very small arteriovenous oxygen difference (about 0.5 ml O_2 per 100 ml blood) even though it has one of the highest metabolic rates in the body. Certain drugs and enzyme poisons that block the cytochrome chain or the formation of ATP stimulate the carotid body. For example both cyanide and dinitrophenol stimulate the carotid body; this may be related to the stimulatory effect of hypoxia on the arterial chemoreceptors. Some studies have shown two different types of cytochrome a_3, with different oxygen affinities, to be present in the carotid body. Ganglionic stimulators such as nicotine also stimulate the carotid body.

Central Chemoreceptors

The central chemoreceptors are exposed to the cerebrospinal fluid and are not in direct contact with the arterial blood. The cerebrospinal fluid is separated from the arterial blood by the blood-brain barrier. Carbon dioxide can easily diffuse through the blood-brain barrier, but hydrogen ions and bicarbonate ions do not. Because of this, alterations in the arterial P_{CO_2} are rapidly transmitted to the cerebrospinal fluid, with a time constant of about 60 s. Changes in arterial pH that are not caused by changes in P_{CO_2} take much longer to influence the cerebrospinal fluid; in fact, the cerebrospinal fluid may have changes in hydrogen ion concentration *opposite* to those seen in the blood in certain circumstances, as will be discussed later in this chapter.

The composition of the cerebrospinal fluid is considerably different from that of the blood. It is formed mainly in the choroid plexus of the lateral ventricles. Enzymes, including carbonic anhydrase, play a large role in cerebrospinal fluid formation: The cerebrospinal fluid is not merely an ultrafiltrate of the plasma. The pH of the cerebrospinal fluid is normally about 7.32, as compared with the pH of 7.40 of arterial blood. The bicarbonate ion concentration is only 3 or 4 meq/liter less than that of the arterial blood, with a mean of about 24 meq/liter, but the P_{CO_2} of the cerebrospinal fluid is about 50 torr. The concentration of proteins in the cerebrospinal fluid is only in the range of 15 to 45 *milligrams*/100 ml whereas the concentration of proteins in the *plasma* normally ranges from 6.6 to 8.6 *grams*/100 ml. This does not even include the hemoglobin in the erythrocytes. Bicarbonate is the only buffer of consequence in the cerebrospinal fluid. Arterial hypercapnia will therefore lead to greater changes in cerebrospinal fluid hydrogen ion concentration than it does in the arterial blood. The brain produces carbon dioxide as an end product of metabolism. Brain carbon dioxide levels are higher than those of the arterial blood, which explains the high P_{CO_2} of the cerebrospinal fluid.

Although the central chemoreceptor was originally believed to be identical with the respiratory neurons in the medullary respiratory center, studies have shown that these cells are not directly responsible for the central ventilatory response to hypercapnia. In fact, the activity of the medullary respiratory center is *depressed* by local hypercapnia. Specialized chemosensitive cells located ventrolaterally at or just beneath the surface of the medulla are generally believed to initiate the response. Increases in their activity are believed to stimulate the medullary respiratory neurons in a manner similar to that of the peripheral chemoreceptors.

The central chemoreceptors respond to local increases in hydrogen ion concentration, P_{CO_2}, or both. They do not respond to hypoxia.

The relative contribution of the peripheral and central chemoreceptors in the ventilatory response to elevated carbon dioxide levels is somewhat controversial. Animals experimentally deprived of the afferent fibers from the arterial chemoreceptors, or patients with surgically removed carotid bodies, show about 80 to 90 percent of the normal total *steady-state* response to elevated inspired carbon dioxide concentrations delivered in hyperoxic gas mixtures, indicating that the peripheral chemoreceptors contribute only 10 to 20 percent of the steady-state response. Other studies performed on normoxic men indicate that up to one-third or one-half of the *onset* of the response can come from the arterial chemoreceptors when rapid changes in arterial P_{CO_2} are made. That is, the central chemoreceptors may be almost solely responsible for establishing the resting ventilatory level or the long-term response to carbon dioxide inhalation, but the peripheral chemoreceptors may be very important in short-term transient responses to carbon dioxide.

Recent investigations have implicated other sensors for carbon dioxide in the body that may influence the control of ventilation. Chemoreceptors within the lungs or airways have been proposed but have not as yet been substantiated or localized.

THE RESPONSE TO HYDROGEN IONS

Ventilation increases nearly linearly with changes in hydrogen ion concentration over the range of 20 to 60 neq/liter. A metabolic acidosis of nonbrain origin results in hyperpnea coming almost entirely from the peripheral chemoreceptors. Hydrogen ions cross the blood-brain barrier too slowly to affect the central chemoreceptor initially. Acidotic stimulation of the peripheral chemoreceptors increases alveolar ventilation and the arterial P_{CO_2} falls. Because the cerebrospinal fluid P_{CO_2} is in a sort of dynamic equilibrium with the arterial P_{CO_2}, carbon dioxide diffuses out of the cerebrospinal fluid and the pH of the cerebrospinal fluid *increases,* thus decreasing stimulation of the central chemoreceptor. If the situation lasts a long time (hours to days), the bicarbonate concentration of the cerebrospinal fluid falls slowly, returning the pH of the cerebrospinal fluid toward the normal 7.32. The mechanism by which this occurs is not completely agreed on. It may represent the slow diffusion of bicarbonate ions across the blood-brain barrier, active transport of bicarbonate ions out of the cerebrospinal fluid, or decreased *formation* of bicarbonate ions by carbonic anhydrase as the cerebrospinal fluid is formed.

Similar mechanisms must alter the bicarbonate concentration in the cerebrospinal fluid in the chronic respiratory acidosis of chronic obstructive lung disease because the pH of the cerebrospinal fluid is nearly normal. In this case the cerebrospinal fluid concentration of bicarbonate increases nearly proportionately to its increased concentration of carbon dioxide.

THE RESPONSE TO HYPOXIA

The ventilatory response to hypoxia arises solely from the peripheral chemoreceptors. The carotid bodies are much more important in this response than are the aortic bodies, which are not capable of sustaining the ventilatory response to hypoxia by themselves. In the absence of the peripheral chemoreceptors, the effect of increasing degrees of hypoxia is a progressive direct *depression* of the central respiratory controller. Therefore, when the peripheral chemoreceptors are intact, their excitatory influence on the central respiratory controller must offset the direct depressant effect of hypoxia.

The response of the respiratory system to hypoxia is shown in Fig.

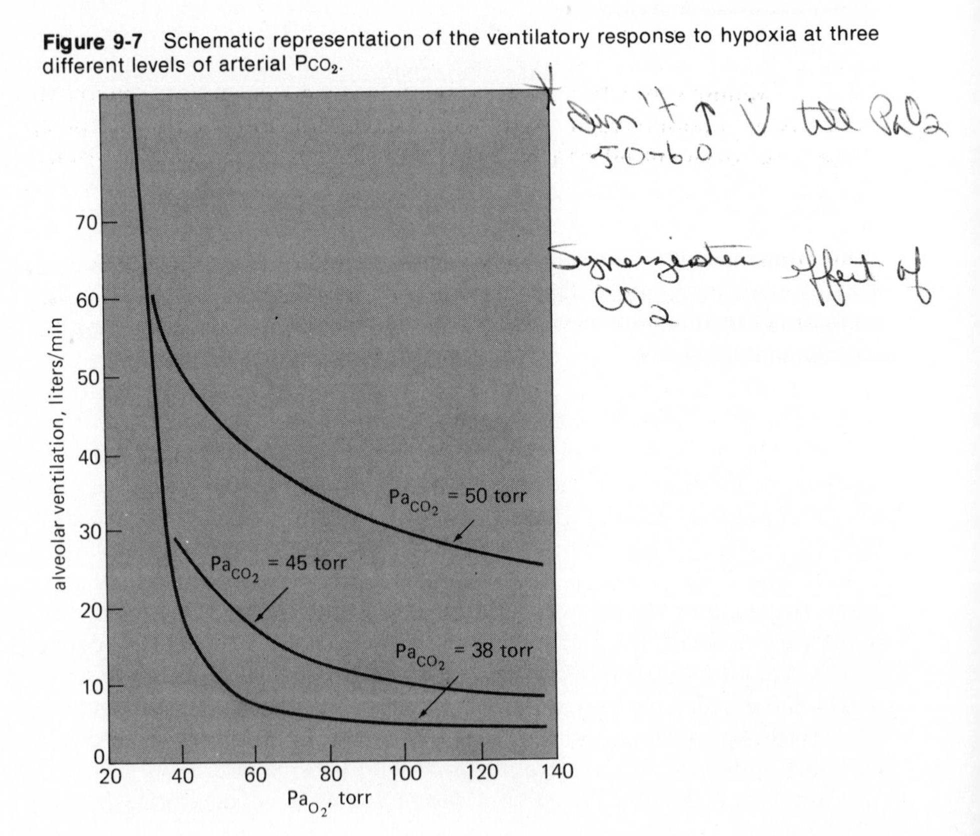

Figure 9-7 Schematic representation of the ventilatory response to hypoxia at three different levels of arterial P_{CO_2}.

9-7. The figure shows that at a normal arterial P_{CO_2} of about 38 to 40 torr, there is very little increase in ventilation until the arterial P_{O_2} falls below about 50 to 60 torr. As expected, the response to hypoxia is potentiated at higher arterial P_{CO_2}'s.

Experiments have shown that the respiratory response to hypoxia is related to the change in *P_{O_2}* rather than the change in *oxygen content*. Therefore anemia (without acidosis) does not stimulate ventilation because the arterial P_{O_2} is normal and the arterial chemoreceptors are not stimulated. Nevertheless, if the arterial oxygen hemoglobin saturation (the Sa_{O_2}) is plotted against ventilation, as seen at left in Fig. 9-8, the response is linear. The reason for this is probably that the curvilinearity of the oxyhemoglobin dissociation curve complements the curvilinearity of the Pa_{O_2} versus ventilation curve.

Hypoxia alone, by stimulating alveolar ventilation, causes a decrease in arterial P_{CO_2}, which may lead to respiratory alkalosis. This will be discussed in the section on altitude in Chap. 11.

THE RESPONSE TO EXERCISE

Exercise increases oxygen consumption and carbon dioxide production; the ventilatory control system must adjust to meet these increased demands. Minute ventilation increases with the level of exercise; it increases linearly with both oxygen consumption and carbon dioxide production up to a level of about 60 percent of the subject's maximal work capacity. Above that level, minute ventilation increases faster than oxygen con-

Figure 9-8 Mean hypoxic response of four normal men at a constant P_{CO_2}. *Left:* Arterial oxygen saturation plotted against ventilation. *Right:* Arterial P_{O_2} plotted against ventilation. [*From Berger, 1977. Reprinted by permission of the N. Engl. J. Med.*].

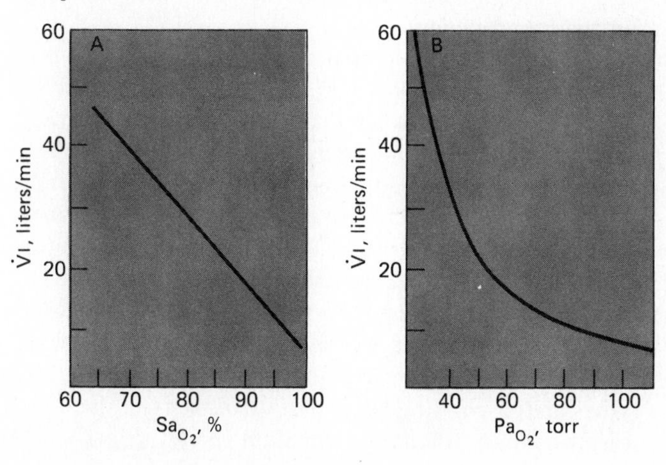

sumption but continues to rise proportionally to the increase in carbon dioxide production. This increase in ventilation above oxygen consumption at high work levels is caused by the increased lactic acid production that occurs as a result of anaerobic metabolism. The hydrogen ions liberated in this process can stimulate the arterial chemoreceptors directly; the buffering of hydrogen ions by bicarbonate ions also results in production of carbon dioxide in addition to that derived from aerobic metabolism.

The ventilatory response to constant work-rate exercise consists of three phases. At the beginning of exercise there is an *immediate* increase in ventilation. This is followed by a phase of slowly increasing ventilation, ultimately rising to a final steady-state phase if the exercise is not too severe. The initial immediate increase in ventilation may constitute as much as 50 percent of the total steady-state response, although it is usually a smaller fraction of the total.

The increase in minute ventilation is usually a result of increases in both tidal volume and breathing frequency. Initially the tidal volume increases more than the rate, but as metabolic acidosis develops the increase in breathing frequency predominates.

The mechanisms by which exercise increases minute ventilation remain controversial. No single factor can fully account for the ventilatory response to exercise and much of the response is unexplained. The immediate increase in ventilation occurs too quickly to be a response to alterations in metabolism or changes in the blood gases. This "neural component" may be partly accounted for by a conditioned reflex, i.e., a learned response to exercise. Experiments have also demonstrated that input to the respiratory centers from proprioceptors located in the joints and muscles of the exercising limbs may play a large role in the ventilatory response to exercise. Passive movements of the limbs of anesthetized animals cause an increase in ventilation. The ventilatory and cardiovascular responses to exercise may be coordinated (and in part initiated) in an "exercise center" in the hypothalamus.

The arterial chemoreceptors do not appear to play a role in the initial immediate ventilatory response to exercise. In mild or moderate exercise (that below the point at which anaerobic metabolism plays a role in energy supply) mean arterial P_{CO_2} and P_{O_2} remain relatively constant, even during the increasing ventilation phase (the "humoral component") and may actually improve. It is therefore unlikely that hypercapnic or hypoxic stimulation of the arterial chemoreceptors is important in the ventilatory response to exercise in this situation. Nevertheless, patients who have had their carotid bodies surgically removed for medical reasons do show a slower increase in ventilation during the second phase of constant work-rate exercise, even in the absence of lactic acidosis. It is possible that the arterial chemoreceptors are responding to greater *oscillations* in the blood gases during exercise, despite relatively constant mean Pa_{CO_2}'s and

Pa_{O_2}'s. During exercise levels above the "anaerobic threshold" these patients do not further increase their ventilation despite metabolic acidosis, indicating the importance of the peripheral chemoreceptors in this portion of the response.

Several investigators have suggested that there may be receptors in the pulmonary circulation that could respond to an increased carbon dioxide load in the mixed venous blood. Others have proposed that receptors in the exercising muscles may send information on the increased muscle metabolism to the respiratory controllers. Thus far, these "mixed venous chemoreceptors" and "metaboreceptors" have not been demonstrated conclusively. The increase in body temperature that occurs during exercise may also contribute to the ventilatory response.

Chapter 10

Nonrespiratory Functions of the Lung

OBJECTIVES

The student understands the nonrespiratory functions of the components of the respiratory system.

1 Lists and describes the mechanisms by which the lung is protected from the contaminants in inspired air.
2 Describes the "air-conditioning" function of the upper airways.
3 Describes the filtration and removal of particles from the inspired air.
4 Describes the removal of biologically active material from the inspired air.
5 Describes the reservoir and filtration functions of the pulmonary circulation.
6 Lists the metabolic functions of the lung, including the handling of vasoactive materials in the blood.

The main function of the respiratory system in general and the lung in particular is gas exchange. The lung does, however, have several other tasks to perform. These nonrespiratory functions of the lung include its own defense against inspired particulate matter, the storage and filtration of blood for the systemic circulation, the handling of vasoactive substances in the blood, and the formation and release of substances used in the alveoli or circulation.

PULMONARY DEFENSE MECHANISMS

Every day about 10,000 liters of air is inspired into the lungs, bringing the air into contact with approximately 50 to 70 m^2 of what may be the most delicate tissues of the body. This inspired air contains (or may contain) dust, pollen, ash, and other products of combustion, microorganisms such as bacteria, particles of substances such as asbestos or silica, and hazardous chemicals or toxic gases. As one reviewer (Green) put it, "each day a surface as large as a tennis court is exposed to a volume of air and contaminants that would fill a swimming pool." In this section we will discuss the mechanisms by which the lungs are protected from contaminants in inspired air.

"Air Conditioning"

The temperature and humidity of the ambient air varies widely and the alveoli must be protected from the cold and from drying out. The mucosa of the nose, the nasal turbinates, the oropharynx, and nasopharynx have a rich blood supply and constitute a large surface area. The nasal turbinates alone have a surface area said to be about 160 cm^2. As inspired air passes through these areas, it is heated to body temperature and humidified, especially if one is breathing through the nose.

Olfaction

Because the olfactory receptors are located in the posterior nasal cavity, rather than in the trachea or alveoli, a person can *sniff* to attempt to detect potentially hazardous gases or dangerous material in the inspired air. This rapid, shallow inspiration brings gases into contact with the olfactory sensors without bringing them into the lung.

Filtration and Removal of Inspired Particles

The respiratory tract has a highly elaborate system for the filtration and removal of particulate matter in the inspired air. The system works better if one is breathing through the nose.

Filtration of Inspired Air Inhaled particles may be deposited in the respiratory tract as a result of impaction, sedimentation, Brownian motion, and other, less important mechanisms. Air passing through the nose is first filtered by passing through the nasal hairs, or *vibrissae*. This removes most particles larger than 10 to 15 μm in diameter. Most of the particles greater than 10 μm in diameter are removed by impacting in the

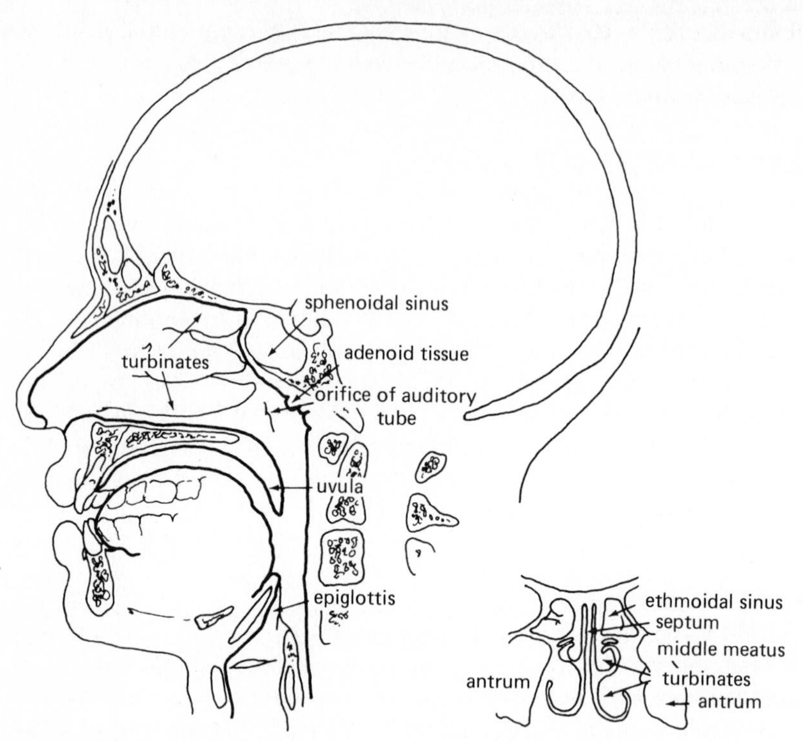

Figure 10-1 Schematic drawing of the upper airways. (*After Proctor, 1964. Reproduced with permission.*)

large surface area of the nasal septum and turbinates, shown in Fig. 10-1. The inspired air stream changes direction abruptly at the nasopharynx and most particles of about 10 μm in diameter impact on the posterior wall of the pharynx because of their inertia. The tonsils and adenoids are located near this impaction site, providing immunologic defense against biologically active material filtered at this point. Air entering the trachea contains few particles larger than 10 μm and most of these will impact mainly at the carina or within the bronchi.

Sedimentation of most particles in the size range of 0.2 to 5 μm occurs by gravity in the smaller airways, where airflow rates are extremely low. Thus most of the particles between 2 and 10 μm in diameter are removed by impaction or sedimentation and become trapped in the mucus lining the upper airways, trachea, bronchi, and bronchioles. Smaller particles and all foreign gases reach the alveolar ducts and alveoli. Some smaller particles (0.1 μm and smaller) are deposited as a result of Brownian motion due to their bombardment by gas molecules. The other particles, between 0.1 and 0.5 μm in diameter, mainly stay suspended as aerosols, and about 80 percent of them are exhaled.

Removal of Filtered Material Filtered material trapped in the mucus lining the respiratory tract can be removed in several ways.

Reflexes in the Airways Mechanical or chemical stimulation of receptors in the nose, trachea, larynx, or elsewhere in the respiratory tract may produce bronchoconstriction, to prevent deeper penetration of the irritant into the airways, and may also produce a cough or a sneeze. A sneeze results from stimulation of receptors in the nose or nasopharynx; a cough results from stimulation of receptors in the trachea. In either case a deep inspiration, often to near the total lung capacity, is followed by a forced expiration against a closed glottis. Intrapleural pressure may rise to more than 100 mmHg during this phase of the reflex. The glottis opens suddenly and pressure in the airways falls rapidly, resulting in compression of the airways and an explosive expiration, with linear airflow velocities said to approach the speed of sound. Such high airflow rates through the narrowed airways are likely to carry the irritant, along with some mucus, out of the respiratory tract. In a sneeze, of course, the expiration is via the nose; in a cough, the expiration is via the mouth. The cough or sneeze reflex is also useful in helping to move the mucus lining of the airways toward the nose or mouth.

Tracheobronchial Secretions and Mucociliary Transport: The "Mucociliary Escalator" The entire respiratory tract, from the upper airways down to the terminal bronchioles, is lined by a mucus-covered ciliated epithelium. The only exceptions are parts of the pharynx and the anterior third of the nasal cavity. A typical portion of the epithelium of the airways (without the layer of mucus that would normally cover it) is shown in Fig. 10-2.

The airway secretions are produced by goblet cells and mucus-secreting glands. The mucus is a complex polymer of mucopolysaccharides. The mucus glands are found mainly in the submucosa near the supporting cartilage of the larger airways. In pathological states, such as chronic bronchitis, the number of goblet cells may increase and the mucus glands may hypertrophy, resulting in greatly increased mucus gland secretion and increased mucus viscosity.

The cilia lining the airways beat in such a way that the mucus covering them is always moved up the airway, away from the alveoli and toward the pharynx. Exactly how the ciliary beating is coordinated is unknown—the cilia do not appear to beat synchronously but probably produce local waves. The mucus blanket appears to be involved in the mechanical linkage between the cilia. The cilia beat at frequencies between 1000 to 1500 beats per minute and the mucus moves progressively faster as it travels from the periphery. In small airways (1 to 2 mm in diameter), linear velocities range from 0.5 to 1 mm/min; in the trachea and bronchi, linear velocities range from 5 to 20 mm/min. Several studies have shown that ciliary function is inhibited or impaired by cigarette smoke.

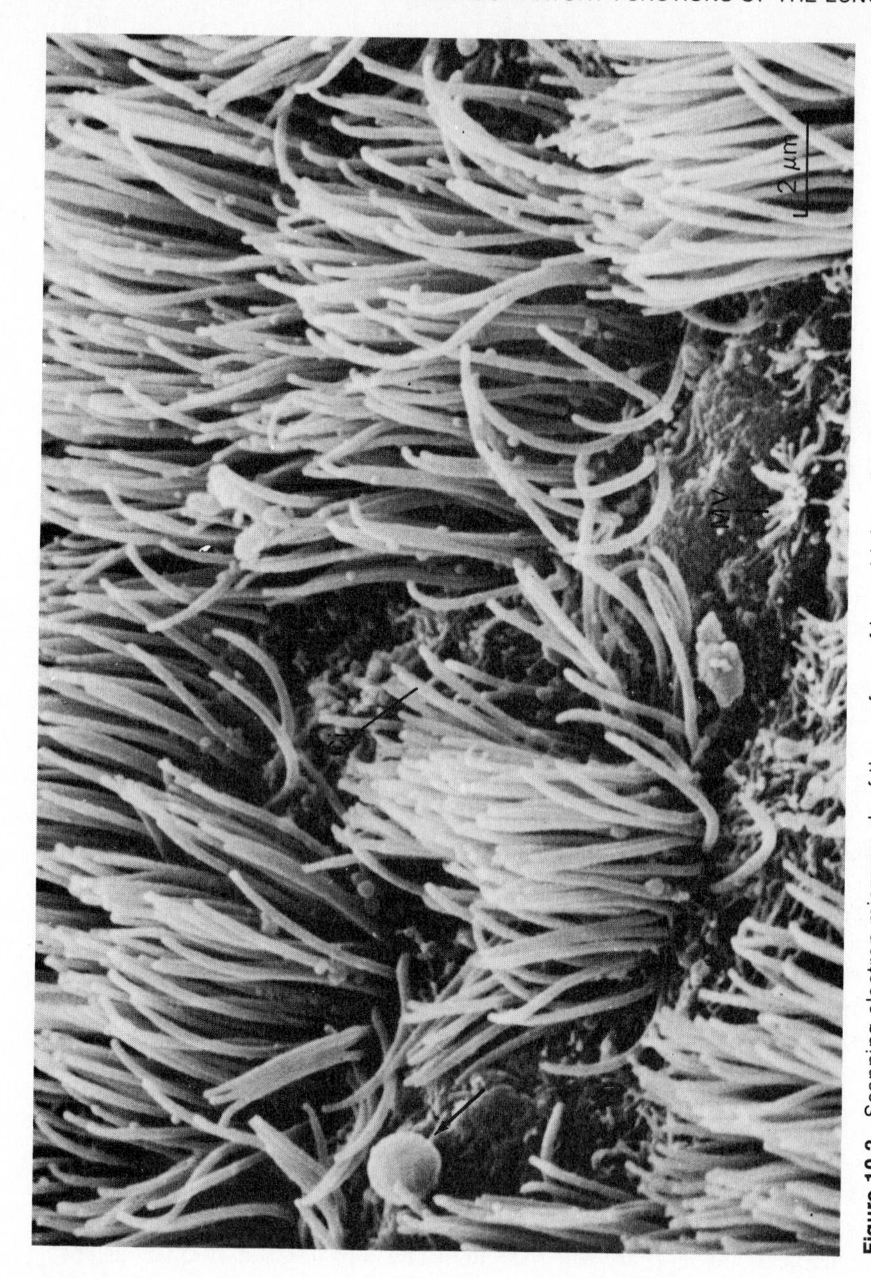

Figure 10-2 Scanning electron micrograph of the surface of bronchiolar epithelium. Ci = cilia; MV = microvilli on surface of unciliated cell. Arrow points to a secretion droplet. (*From Weibel, 1980.*)

The "mucociliary escalator" is an especially important mechanism for the removal of inhaled particles that come to rest in the airways. Material trapped in the mucus is continuously moved upward toward the pharynx. This movement can be greatly increased during a cough, as described previously. Mucus that reaches the pharynx is usually swallowed, expectorated, or removed by blowing one's nose. It is important to remember that patients who cannot clear their tracheobronchial secretions (an intubated patient or a patient who cannot cough adequately) continue to produce secretions. If the secretions are not removed from the patient by suction or other means, airway obstruction will develop.

Defense Mechanisms of the Terminal Respiratory Units Inspired material that reaches the terminal airways and alveoli may be removed in several ways, including ingestion by alveolar macrophages, nonspecific enzymatic destruction, entrance into the lymphatics, and immunologic reactions.

Alveolar Macrophages Alveolar macrophages are large mononuclear ameboid cells that inhabit the alveolar surface. Inhaled particles engulfed by alveolar macrophages may be destroyed by their lysosomes. Most bacteria are digested in this manner. Some material ingested by the macrophages, however, such as silica, is not degradable by the macrophages and may even be toxic to them. If the macrophages carrying such material are not removed from the lung, the material will be redeposited in the alveolar surface on the death of the macrophages. The mean life span of alveolar macrophages is believed to be 1 to 5 weeks. The main exit route of macrophages carrying such nondigestible material is migration to the mucociliary escalator via the pores of Kohn and eventual removal through the airways. Particle-containing macrophages may also migrate from the alveolar surface into the septal interstitium, from which they may enter the lymphatic system or enter the mucociliary escalator. Macrophage function has been shown to be inhibited by cigarette smoke. A scanning electron micrograph of an alveolar macrophage is shown in Fig. 10-3.

Other Methods of Particle Removal or Destruction Some particles reach the mucociliary escalator because the alveolar fluid lining itself is slowly moving upward toward the respiratory bronchioles. Others penetrate into the interstitial space or enter the blood, where they are phagocytized by interstitial macrophages or blood phagocytes or enter the lymphatics. Particles may be destroyed or detoxified by surface enzymes and factors in the serum and in airway secretions. These include *lysozymes,* found mainly in leukocytes and known to have bactericidal properties; *lactoferrin,* which is synthesized by polymorphonuclear lymphocytes and by glandular mucosal cells and is a potent bacteriostatic agent; *alpha*$_1$ *antitrypsin,* which inactivates proteolytic enzymes released from bacteria or from cells involved in defense of the lung; *interferon,* a

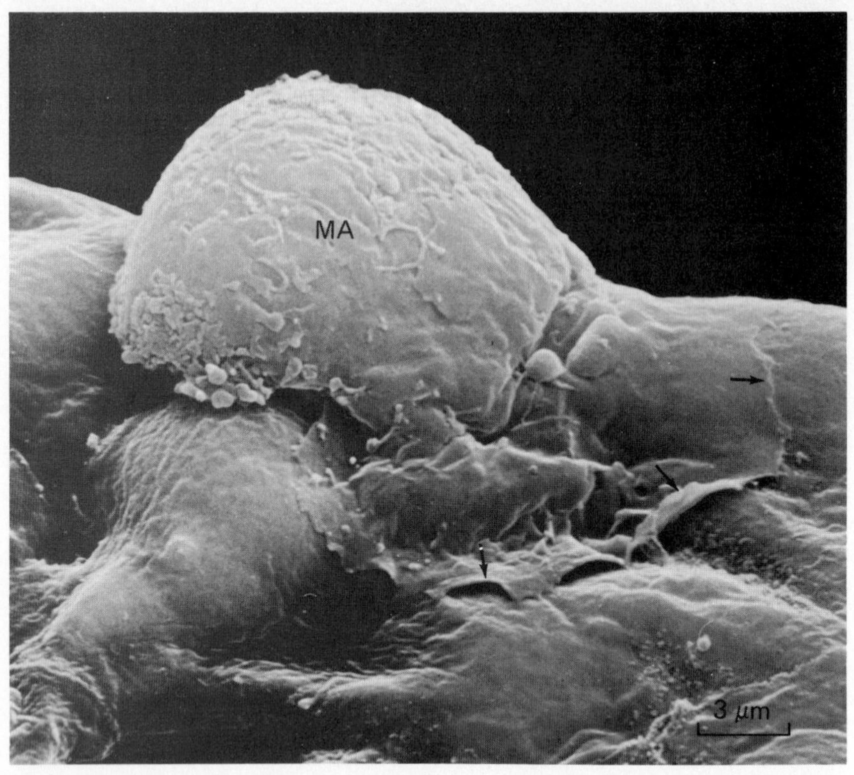

Figure 10-3 Scanning electron micrograph of an alveolar macrophage on the epithelial surface of a human lung. Arrows point to the advancing edge of the cell. (*From Weibel, 1980.*)

potent antiviral substance that may be produced by macrophages and lymphocytes; and *complement,* which participates as a cofactor in antigen-antibody reactions and may also participate in other aspects of cellular defense. Finally, many biologically active contaminants of the inspired air may be removed by antibody-mediated or cell-mediated immunologic responses. A schematic diagram summarizing bronchoalveolar pulmonary defense mechanisms is shown in Fig. 10-4.

NONRESPIRATORY FUNCTIONS OF THE PULMONARY CIRCULATION

The pulmonary circulation, strategically located between the systemic veins and arteries, is well-suited for several functions not directly related to gas exchange. The entire cardiac output passes through the very large surface area of the pulmonary capillary bed, allowing the lungs to act as a site of blood filtration and storage, as well as a site for the metabolism of vasoactive constituents of the blood.

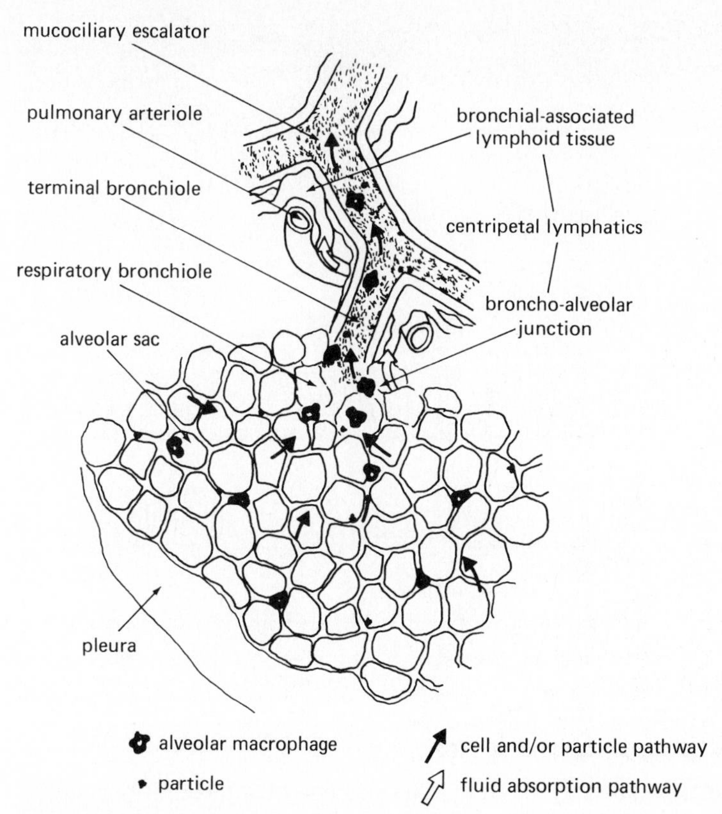

Figure 10-4 Schematic diagram of bronchiolar-alveolar defense mechanisms. (*After Green, 1977. Reproduced with permission.*)

Reservoir for the Left Ventricle

The pulmonary circulation, because of its high compliance and the negative intrapleural pressure, contains about 250 to 300 ml of blood per square meter of body surface area. This would give a typical adult male a pulmonary blood volume of about 500 ml. This large blood volume allows the pulmonary circulation to act as a reservoir for the left ventricle. If left ventricular output is transiently greater than systemic venous return, left ventricular output can be maintained for a few strokes by drawing on blood stored in the pulmonary circulation. This can be demonstrated in an experiment such as that depicted in Fig. 10-5.

In this experiment a balloon was inflated in the main pulmonary artery. Pulmonary blood flow dropped to nearly zero almost immediately and pressure in the superior vena cava rose dramatically. Note that aortic flow and systemic arterial blood pressure were maintained for two or three strokes of the left ventricle, even in this instance where virtually no blood entered the pulmonary circulation.

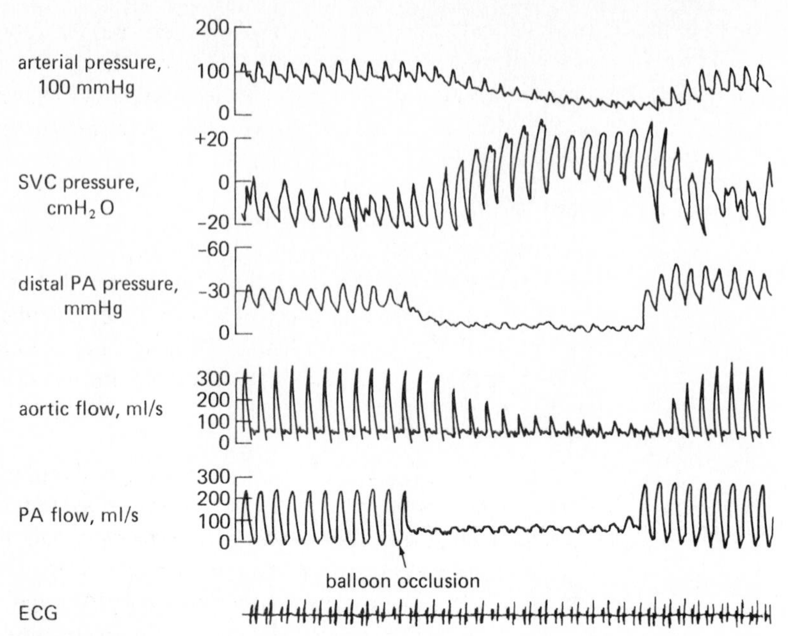

Figure 10-5 Effects of balloon occlusion of the main pulmonary artery. (*From Hoffman, 1965.*)

The Pulmonary Circulation as a Filter

Because virtually all mixed venous blood must pass through the pulmonary capillaries, the pulmonary circulation acts as a filter, protecting the systemic circulation from materials that enter the blood. The particles filtered, which may enter the circulation as a result of natural processes, trauma, or therapeutic measures, may include small fibrin or blood clots, fat cells, bone marrow, detached cancer cells, gas bubbles, agglutinated erythrocytes (especially in sickle cell anemia), masses of platelets or leukocytes, and debris from stored blood or intravenous solutions. If these particles were to enter the arterial side of the systemic circulation they might occlude vascular beds with no other source of blood flow. This occlusion would be particularly disastrous if it occurred in the blood supply to the central nervous system or the heart.

The lung can perform this very valuable service because there are many more pulmonary capillaries present in the lung than are necessary for gas exchange at rest. Obviously, no gas exchange can occur distal to a particle embedded in and obstructing a capillary, and so this mechanism is limited by the ability of the lung to remove such filtered material. If particles are experimentally suspended in venous blood and are then trapped in the pulmonary circulation, the diffusing capacity usually decreases for 4 to 5 days and then returns to normal. The mechanisms for removal of material trapped in the pulmonary capillary bed include lytic enzymes in the capillary endothelium, ingestion by macrophages, and penetration to the lymphatic system. Patients on cardiopulmonary bypass do not have the benefit of this pulmonary capillary filtration and blood administered to these patients must be filtered for them.

Fluid Exchange and Drug Absorption

The low pulmonary capillary hydrostatic pressure, normally far less than the colloid osmotic pressure of the plasma proteins, tends to pull fluid from the alveoli into the pulmonary capillaries and keep the alveolar surface free of liquids other than pulmonary surfactant. Water taken into the lungs is rapidly absorbed into the blood. This protects the gas exchange function of the lungs and opposes transudation of fluid from the capillaries to the alveoli.

Drugs or chemical substances that readily pass through the alveolar-capillary barrier by diffusion or other means rapidly enter the systemic circulation. The lungs are frequently used as a route of administration of anesthetic gases such as halothane and nitrous oxide. Aerosol drugs intended for the airways only, such as the bronchodilator isoproterenol, may rapidly pass into the systemic circulation, where they may have large effects. The effects of isoproterenol, for example, could include cardiac stimulation and vasodilation.

METABOLIC FUNCTIONS OF THE LUNG

Until recently the lung was thought of as an organ with little metabolic activity. The main function of the lung, gas exchange, is accomplished by passive diffusion. Movement of air and blood are accomplished by the muscles of respiration and the right ventricle. Because the lung appeared to do little that required energy and did not appear to produce any substances utilized elsewhere in the body, it was not believed to have any metabolic requirements other than those necessary for the maintenance of its own cells. In the last two decades, however, the metabolic activities of the lung have become an area of intense investigation, and the lung has been shown to be involved in the conversion or uptake of vasoactive substances found in mixed venous blood and the production, storage, and release of substances used locally in the lung or elsewhere in the body.

Metabolism of Vasoactive Substances

Many vasoactive substances are inactivated, altered, or removed from the blood as they pass through the lungs. The site of this metabolic activity is believed to be the endothelium of the vessels of the pulmonary circulation, which constitute a tremendous surface area in contact with the mixed venous blood. For example, as shown on Table 10-1, prostaglandins E_1,

Table 10-1 Uptake or Conversion by the Lungs of Chemical Substrates in Mixed Venous Blood

Substance in mixed venous blood	Result of a single pass through the lung
Prostaglandins E_1, E_2, $F_{2\alpha}$	Almost completely removed
Prostaglandins A_1, A_2	Not affected
Serotonin	85–95% removed
Acetylcholine	Inactivated by cholinesterases in blood
Histamine	Not affected
Epinephrine	Not affected
Norepinephrine	Approximately 30% removed
Isoproterenol	Not affected
Dopamine	Not affected
Bradykinin	Approximately 80% inactivated
Angiotensin I	Approximately 70% converted to angiotensin II
Angiotensin II	Not affected
Vasopressin	Not affected
Oxytocin	Not affected
Gastrin	Not affected
ATP, AMP	40–90% removed

E_2, and $F_{2\alpha}$ are nearly completely removed in a single pass through the lungs. On the other hand, prostaglandins A_1 and A_2 are not affected by the pulmonary circulation. Similarly, about 30 percent of the norepinephrine in mixed venous blood is removed by the lung, but epinephrine and isoproterenol are unaffected.

These alterations of vasoactive substances in the lung imply several things. First, some substances released into specific vascular beds for *local* effects are inactivated or removed as they pass through the lungs, preventing them from entering the systemic circulation. Other substances, apparently intended for more general effects, are not affected. Second, in the case of those substances that are affected by passing through the lungs, there may be profound differences in the response of a patient receiving an injection or infusion of one of these substances, depending on whether it is administered via an arterial or venous catheter.

Formation and Release of Chemical Substances for Local Use

Several substances that produce effects in the lung have been shown to be synthesized and released by pulmonary cells. The most familiar of these is *pulmonary surfactant,* which is synthesized in type II alveolar epithelial cells and released into the alveolar surface. Surfactant plays an important role in reducing the alveolar elastic recoil due to surface tension and in stabilizing the alveoli, as discussed in Chap. 2. *Histamine* and *serotonin* can be released from mast cells in the lung in response to pulmonary embolism or anaphylaxis and cause bronchoconstriction and may initiate other cardiopulmonary reflexes. Whatever substance or substances are the *chemical mediators* involved in hypoxic pulmonary vasoconstriction, discussed in Chap. 4, are produced in and act in the lung. Although histamine and serotonin, as well as various prostaglandins, have been considered as the mediator at one time or another, there is currently no consensus as to what the mediator is or even if there is a mediator involved in this response. Many substances are also produced by cells of the lung and released into the alveoli and airways, including mucus and other tracheobronchial secretions, the surface enzymes, proteins and other factors, and immunologically active substances discussed earlier in this chapter. These are produced by goblet cells, submucosal gland cells, Clara cells, and macrophages.

Formation and Release into the Blood of Substances Produced by Lung Cells

Bradykinin, histamine, serotonin, heparin, and prostaglandins E_2 and $F_{2\alpha}$ are all stored in the lung and may be released into the general circulation under various circumstances. For example, heparin, histamine, serotonin, and prostaglandins E_2 and $F_{2\alpha}$ are released during anaphylactic shock.

Other Metabolic Functions

The lung must be able to meet its own cellular energy requirements as well as be able to respond to injury. The type II alveolar epithelial cell also plays a major role in the response of the lung to injury. As type I alveolar epithelial cells are injured, type II cells proliferate to reestablish a continuous epithelial surface. Studies in animals have shown that these type II cells can develop into type I cells after injury.

Chapter 11

The Respiratory System Under Stress

OBJECTIVES

The student uses the knowledge he or she gained from the preceding chapters of this book to predict the response of the respiratory system to three physiological stresses: exercise, ascent to altitude, and diving.

1 Identifies the physiological stresses involved in exercise.
2 Predicts the responses of the respiratory system to acute exercise.
3 Describes the effects of long-term exercise programs (training) on the respiratory system.
4 Identifies the physiological stresses involved in the ascent to altitude.
5 Predicts the initial responses of the respiratory system to the ascent to altitude.
6 Describes the acclimatization of the cardiovascular and respiratory systems to residence at high altitudes.
7 Identifies the physiological stresses involved in diving.
8 Predicts the responses of the respiratory system to various types of diving.

This chapter is mainly intended to be a review of the preceding chapters of the book. The responses of the respiratory system to three physiological stresses are examined as they relate to the material already covered; the discussions of the responses to each stress will therefore be brief and rather superficial. For a more complete discussion of each stress, consult the bibliography for this chapter.

EXERCISE AND THE RESPIRATORY SYSTEM

Exercise increases the metabolism of the appropriate muscles. It stresses the respiratory system by increasing the demand for oxygen and by increasing the production of carbon dioxide. Moderate-to-severe levels of exercise also cause increased lactic acid production. The respiratory and cardiovascular systems must increase the volume of oxygen supplied to the exercising tissues and increase the removal of carbon dioxide and hydrogen ions from the body.

Acute Effects

The effects of exercise in an untrained person are mainly a function of an increase in the cardiac output coupled with an increase in alveolar ventilation.

Control of Breathing As discussed at the end of Chap. 9, both the tidal volume and the breathing frequency are increased during exercise. The causes of the increased alveolar ventilation during exercise were discussed in that section.

Mechanics of Breathing The work of breathing is increased during exercise. Larger tidal volumes result in increased work necessary to overcome the elastic recoil of the lungs during inspiration because the lungs are less compliant at higher lung volumes. Of course, the greater elastic recoil of the lungs tends to make expiration easier, but this is offset by other factors. The high airflow rates generated during exercise result in a much greater airways resistance component of the work of breathing. Greater turbulence and dynamic compression of airways secondary to active expiration combine to greatly increase the work of breathing. (Recall that during turbulent airflow $\Delta P = \dot{V}^2 R$.) Increasing airflow rates especially increase the resistive work of breathing through the nose: Minute ventilations above about 40 liters/min are normally accomplished by breathing through the mouth.

Alveolar Ventilation In normal adults the resting minute ventilation ($\dot{V}_E$) of 5 to 6 liters/min can be increased to as much as 150 liters/min during short periods of maximal exercise. Maximal increases in cardiac output during exercise are only in the range of four to six times the resting level in healthy adults, compared with this 25-fold potential increase in minute ventilation. Therefore, it is the cardiovascular system rather than the respiratory system that is the limiting factor in exercise by normal healthy people.

As discussed in Chap. 9, at less strenuous levels of exercise, the increase in ventilation is mainly accomplished by increasing the tidal volume. During strenuous exercise, the tidal volume usually increases to a maximum of about 50 to 60 percent of the vital capacity of a normal subject, or about 2.5 to 3.0 liters in an average-sized man. This increase in tidal volume appears to occur mainly at the expense of the inspiratory reserve volume, with the expiratory reserve volume somewhat less affected. An increase in the central blood volume (caused by increased venous return) may decrease the total lung capacity slightly. The residual volume and functional residual capacity may be unchanged or slightly elevated. The vital capacity may be slightly decreased or unchanged. With strenuous exercise the breathing frequency may increase to 40 to 50 breaths per minute in healthy adults (and as high as 70 breaths per minute in children).

The anatomic dead space may increase slightly in inspiration during exercise because of airway distention at high lung volumes; any alveolar dead space present at rest normally decreases as cardiac output increases. As a result there is little change in physiological dead space during exercise. Because the tidal volume does increase, however, the ratio of physiological dead space to tidal volume (V_D/V_T) decreases.

The arterial P_{O_2} stays relatively constant during even strenuous exercise. Arterial P_{CO_2} also stays relatively constant until anaerobic metabolism results in appreciable lactic acid generation. The hydrogen ions liberated directly stimulate alveolar ventilation and may cause arterial P_{CO_2} to fall a few torr below the resting arterial P_{CO_2}.

The regional differences in alveolar ventilation seen in upright lungs (which were discussed in Chap. 3) are probably attenuated during exercise. The very large tidal volumes, occurring at the expense of both the inspiratory and expiratory reserve volumes, indicate that during inspiration alveoli in more dependent regions of the lung are more fully inflated. These alveoli, however, may also suffer airway collapse at higher lung volumes during active expirations. Similarly, alveoli in upper portions of the lung (with respect to gravity) should deflate more fully during expiration, resulting in greater ventilation of upper parts of the lung.

Pulmonary Blood Flow As already mentioned, the cardiac output increases linearly with oxygen consumption during exercise. This normally occurs more as a result of an autonomically mediated increase in heart rate than an increase in stroke volume. An increased venous return, in part due to deeper inspiratory efforts, also contributes to the increase in cardiac output. Mean pulmonary artery and mean left atrial pressures increase, but the increase is not as great as the increase in pulmonary blood flow. This indicates a decrease in pulmonary vascular resistance. As discussed in Chap. 4, this decrease occurs passively, by recruitment and distention of pulmonary vessels. Much of the recruitment of pulmonary blood vessels occurs in upper regions of the lung, thus tending to decrease the regional inhomogeneity of pulmonary blood flow discussed in Chap. 4. The expected effect of the deeper tidal volumes and active expirations that occur during exercise is to increase pulmonary vascular resistance. During active expiration the extraalveolar vessels should be compressed; during inspiration the alveolar vessels should be compressed. The fact that mean pulmonary vascular resistance decreases indicates that the effects of recruitment and distention are greater than those of extravascular compression.

Ventilation-Perfusion Relationships The more uniform regional ventilation and perfusion that occur during exercise result in a much more uniform distribution of $\dot{V}_A/\dot{Q}$'s throughout the lung. Studies done on normal subjects engaged in exercise in the upright position have demonstrated a greatly increased perfusion of upper regions of the lung resulting in nearly uniform $\dot{V}_A/\dot{Q}$'s from the bottom of the lung to the top of the lung. The ventilation-perfusion ratios were close to 1.0 during exercise, with slightly higher (1.2 to 1.4) ratios in the uppermost and lowermost regions. A comparison of these findings with the $\dot{V}_A/\dot{Q}$'s seen at rest in Fig. 5-5 demonstrates the reduced "scatter" of ventilation-perfusion ratios during exercise.

Diffusion through the Alveolar-Capillary Barrier The diffusing capacities for oxygen and carbon dioxide normally increase substantially during exercise. Some studies have shown a nearly linear increase in diffusing capacity as oxygen uptake increases, although the diffusing capacity may reach a maximum level before the $\dot{V}O_2$ does. The increase in diffusing capacity during exercise is largely a result of the increase in pulmonary blood flow. Recruitment of capillaries in upper regions of the lungs increases the surface area available for diffusion. Increased linear velocity of blood flow through pulmonary capillaries reduces the time red blood cells spend in contact with the alveolar air to less than the 0.75 s normally seen at rest. As noted in Chap. 6, the P_{O_2} and P_{CO_2} of the plasma in the pulmonary capillaries normally equilibrate with the alveolar P_{O_2} and P_{CO_2} within about the first 0.25 s of the time blood spends in the pulmonary capillaries. After this equilibration no further gas diffusion between the equilibrated blood and the alveoli takes place because the partial pressure gradient ($P_1 - P_2$ in Fick's law) is equal to zero. Increased velocity of blood flow through the lung therefore increases the diffusing capacity by bringing unequilibrated blood into the lung faster, thus maintaining the partial pressure gradient for diffusion.

Another related factor, less dependent on the increased cardiac output, that helps to increase diffusion during exercise is that the mixed venous P_{O_2} may be lower and the mixed venous P_{CO_2} may be higher than those seen at rest. These factors may also help to increase and maintain the partial pressure gradients for diffusion.

The total effect on diffusion through the alveolar-capillary barrier of the increased surface area and the better maintenance of the alveolar-capillary partial pressure gradients may be seen by reviewing Fick's law for diffusion:

$$\dot{V}_{gas} \propto \frac{A \times D \times (P_1 - P_2)}{T}$$

The thickness of the alveolar-capillary barrier may also be affected during exercise, but the net effect may be either an increase or a decrease. At high lung volumes the alveolar vessels are compressed and the thickness of the barrier may decrease. On the other hand, high cardiac outputs may be associated with vascular congestion, increasing the thickness of the barrier.

Oxygen and Carbon Dioxide Transport by the Blood The loading of carbon dioxide into the blood and the unloading of oxygen from the blood are enhanced in exercising muscles. Oxygen unloading is improved because the P_{O_2} in the exercising muscle is decreased, causing a larger percentage of deoxyhemoglobin. Oxygen unloading is also enhanced by the rightward shift of the oxyhemoglobin dissociation curve caused by the elevated P_{CO_2}'s (the Bohr effect), hydrogen ion concentrations, and temperatures (and possibly 2,3-DPG) found in exercising muscle. Low capillary P_{O_2}'s should also lead to improved CO_2 loading because lower oxyhemoglobin levels shift the carbon dioxide dissociation curve to the left (the Haldane effect).

Acid-Base Balance Exercise severe enough to cause a significant degree of anaerobic metabolism results in metabolic acidosis secondary to the increased lactic acid production. As discussed previously, the hydrogen ions generated by this process stimulate the arterial chemoreceptors (especially the carotid bodies) and stimulate a further compensatory increase in alveolar ventilation, maintaining arterial pH near the normal level.

Training Effects

The ability to perform physical exercise increases with training. Most of the changes that occur as a result of physical training, however, are a function of alterations in the cardiovascular system and in muscle metabolism rather than the respiratory system. The maximal oxygen uptake increases with physical training. This increase appears to be mainly a result of an increased maximal cardiac output. As was stated earlier in this chapter, the maximal cardiac output is probably a limiting factor in exercise. Physical training lowers the resting heart rate and increases the resting stroke volume. The maximal heart rate does not appear to be affected by physical training, but the heart rate of a trained person is lower than that of an untrained person at any level of physical activity. Stroke volume is increased. The arterial hemoglobin concentration and the hematocrit do not appear to change with physical training at sea level, but the arteriovenous oxygen content difference does appear to increase with physical training. This is probably a function of increased effects of local pH, P_{CO_2}, and temperature in the exercising muscles. Blood volume is usually increased by training.

Physical training increases the oxidative capacity of skeletal muscle by inducing mitochondrial proliferation and increasing the concentration of oxidative enzymes and the synthesis of glycogen and triglyceride. These alterations result in lower concentrations of blood lactate in trained subjects than those found in untrained people, reflecting increased aerobic energy production. Nevertheless, blood lactate levels during maximal exercise may be higher in trained athletes than in untrained people.

Maximal ventilation and resting ventilation do not appear to be affected by physical training, but ventilation at submaximal loads is decreased, probably because of the lower lactic acid levels of the trained person during submaximal exercise. The strength and endurance of the respiratory muscles appear to improve with training. Total lung capacity is not affected by training: Vital capacity may be normal or elevated. Pulmonary diffusing capacity is often elevated in athletes, probably as a result of their increased blood volumes and maximal cardiac outputs.

ALTITUDE AND ACCLIMATIZATION

As one ascends to greater altitudes, total barometric pressure decreases; the total barometric pressure at any altitude is proportional to the weight of the air above it. Air is attracted to the earth's surface by gravity. Because air is compressible, the change in barometric pressure per change in vertical distance is not constant. There is a greater change in barometric pressure per change in altitude closer to the earth's surface than there is at very great altitudes.

The inspired Po_2 falls as the total barometric pressure falls, but the decrease is not exactly linear. The fractional concentration of oxygen in the atmosphere does not change appreciably with altitude. Oxygen constitutes about 21 percent of the total pressure of dry ambient air, and so the Po_2 of dry air at any altitude is about 0.21 × total barometric pressure at that altitude. Water vapor pressure, however, must also be considered in calculations of the Po_2. The water vapor pressure depends on the temperature and humidity of the air. As the inspired air passes through the airways, it is warmed to body temperature and completely humidified. Therefore the partial pressure exerted by the water vapor in the air entering the alveoli is fixed at 47 torr.

The alveolar Po_2 can therefore be calculated by using the alveolar air equation discussed in Chap. 3:

$$P_{A_{O_2}} = P_{I_{O_2}} - \frac{P_{A_{CO_2}}}{R} + [F]$$

The inspired Po_2 is equal to 0.21 times the total barometric pressure (if ambient air is breathed) after the subtraction of the water vapor pressure of 47 torr.

$$P_{I_{O_2}} = 0.21 \times (P_{bar} - 47 \text{ torr})$$

The alveolar Pco_2 falls at greater altitudes because hypoxic stimulation of the arterial chemoreceptors increases alveolar ventilation. For example, at an altitude of 15,000 ft the total barometric pressure is about 429 torr. The inspired Po_2 is therefore 0.21 × (429 − 47), or 80.2, torr. The alveolar Pco_2 is likely to be decreased to about 32 torr, resulting in a $P_{A_{O_2}}$ of about 45 torr. At 18,000 ft the total barometric pressure is about 380 torr; at 20,000 ft it is 349 torr. At 50,000 ft the total barometric pressure is only 87 torr. Even if 100% O_2 is breathed the $P_{A_{O_2}}$ and $P_{A_{CO_2}}/R$ can only *total* 40 torr after water vapor pressure is subtracted. At 63,000 ft the total barometric pressure is 47 torr and the fluid in blood "boils."

Acute Effects

An unacclimatized person suffers a deterioration of nervous system function as he or she ascends rapidly to great heights. Similar dysfunctions occur if cabin pressure is lost in an airplane. The symptoms are mainly due to hypoxia and may include sleepiness, laziness, a false sense of well-being, impaired judgment, blunted pain perception, increasing errors on simple tasks, decreased visual acuity, clumsiness, and tremors. Severe hypoxia, of course, may result in a loss of consciousness or even death.

If an unacclimatized person ascends to a moderate altitude, he or she may suffer from a group of symptoms known collectively as *mountain sickness.* The symptoms include headache, dizziness, breathlessness at rest, weakness, nausea, sweating, palpitations, dimness of vision, partial deafness, sleeplessness, and dyspnea on exertion. These symptoms are a result of both hypoxia and hypocapnia and alkalosis.

Control of Breathing The decreased alveolar and arterial P_{O_2}'s that occur at altitude result in stimulation of the arterial chemoreceptors and an increase in alveolar ventilation; the central chemoreceptors are not responsive to hypoxia. At an arterial P_{O_2} of 45 torr, minute ventilation is approximately doubled. Because carbon dioxide production is initially normal (it does increase with the elevated work of breathing caused by greater alveolar ventilation), alveolar and arterial P_{CO_2} fall, causing respiratory alkalosis. Arterial hypocapnia also results in "diffusion" of carbon dioxide out of the cerebrospinal fluid, causing an increase in the pH of the cerebrospinal fluid. The central chemoreceptors are therefore not only unresponsive to the hypoxia of altitude; their activity is depressed by the secondary hypocapnia and alkalosis of the cerebrospinal fluid.

Mechanics of Breathing The increased rate, and especially depth, of breathing increase the work of breathing. Greater transpulmonary pressures are necessary to generate greater tidal volumes and also to overcome the possible effects of vascular engorgement of the lung, which will be discussed later in this chapter. High ventilatory rates may be accompanied by active expiration, resulting in dynamic compression of airways. This airway compression, coupled with a reflex parasympathetic bronchoconstriction in response to the arterial hypoxemia, results in elevated resistance work of breathing. More turbulent airflow, likely to be encountered at elevated ventilatory rates, may also contribute to elevated resistance work. Maximum airflow rates may increase because of decreased gas density.

Alveolar Ventilation The anatomic dead space may decrease slightly at altitude because of the reflex bronchoconstriction or increase slightly because of the opposing effect of increased tidal volumes. In any event the ratio of dead space to tidal volume falls with greater tidal volumes. A more uniform regional distribution of alveolar ventilation is also expected at altitude because of deeper inspirations and fuller expirations. Previously collapsed or poorly ventilated alveoli will be better ventilated.

Pulmonary Blood Flow There is an increase in cardiac output, heart rate, and systemic blood pressure at altitude. These effects are probably a result of increased sympathetic stimulation of the cardiovascular system secondary to arterial chemoreceptor stimulation and increased lung inflation. There may also be a direct stimulatory effect of hypoxia on the myocardium. Alveolar hypoxia results in pulmonary vasoconstriction—hypoxic pulmonary vasoconstriction. The increased cardiac output, along with hypoxic pulmonary vasoconstriction and sympathetic stimulation of larger pulmonary vessels, results in an increase in mean pulmonary artery pressure and tends to abolish any preexisting zone 1 by recruiting previously unperfused capillaries. Undesirable consequences of these effects include vascular distention and engorgement of the lung secondary to the pulmonary hypertension (which may lead to "high altitude pulmonary edema") and a greatly increased right ventricular work load.

Ventilation-Perfusion Relationships The increase in pulmonary blood flow seen acutely at altitude, coupled with the more uniform alveolar ventilation, would be expected to make regional $\dot{V}_A/\dot{Q}$ more uniform and closer to 1.0. Surprisingly, studies have not shown striking differences in ventilation-perfusion relationships at altitude.

Diffusion through the Alveolar-Capillary Barrier At altitude the partial pressure gradient for oxygen diffusion is decreased because the alveolar P_{O_2} is decreased more than the mixed venous P_{O_2}. This decrease in the partial pressure gradient is partly offset by the effects of the increase in cardiac output and increased pulmonary artery pressure, which increase the surface area available for diffusion and decrease the time erythrocytes spend in pulmonary capillaries. The thickness of the barrier may be slightly decreased at higher lung volumes or increased because of pulmonary vascular distention.

Oxygen and Carbon Dioxide Transport by the Blood Oxygen loading in the lung may be compromised at alveolar P_{O_2}'s low enough to be below the flat part of the oxyhemoglobin dissociation curve, causing a low arterial oxygen content. Hypocapnia may aid somewhat in oxygen loading in the lung but will interfere with oxygen unloading at the tissues. The main short-term compensatory mechanism for maintenance of oxygen *delivery* is the increased cardiac output. An area of special difficulty is the cerebral circulation: Hypocapnia is a strong cerebral vasoconstrictor. Thus the brain receives not only blood with a low oxygen content, but also receives a reduced blood flow.

Acid-Base Balance As already mentioned the increased alveolar ventilation at altitude results in hypocapnia and respiratory alkalosis.

Acclimatization

Longer-term compensations to the ascent to high altitude begin to occur after several hours and continue for days or even weeks. The immediate responses to the ascent and the early and late adaptive responses are summarized in Table 11-1.

Table 11-1 Physiological Responses to High Altitude Relative to Sea-level Control Values*

	Immediate	Early adaptive (72 h)	Late adaptive (2 to 6 weeks)
Spontaneous ventilation:			
Minute ventilation	↑	↑	↑
Respiratory rate	Variable	Variable	Variable
Tidal volume	↑	↑	↑
Arterial P_{O_2}	↓	↓	↓
Arterial P_{CO_2}	↓	↓	↓
Arterial pH	↑	↑↔	↑↔
Arterial HCO_3	↔	↓	↓
Evaluation of lung function			
Vital capacity	↔	↔	↔
Maximum airflow rates	↑	↑	↑
Functional residual capacity	↔	↔	↔
Ventilatory response to inhaled CO_2	↔	↑	↑
Ventilatory response to hypoxia	↔	↔	↔
Pulmonary vascular resistance	↑	↑	↑
Oxygen transport			
Hemoglobin	↔	↑	↑
Erythropoietin	↑	↔	↔
P_{50}	↓	↑	↑
2,3-DPG	↔	↑	↑
Cardiac output	↑	↔	↔↓
Central nervous system			
Headaches, nausea, insomnia	↑	↔	↔
Perception, judgment	↓	↔	↔
Spinal fluid pH	↑	↔	↔
Spinal fluid HCO_3^-	↔	↓	↓

* These values apply to native sea-level inhabitants.
Source: Reproduced with permission from Guenter, 1976.

Renal compensation for respiratory alkalosis begins within a day: Renal excretion of base is increased and hydrogen ions are conserved. A second major compensatory mechanism is erythropoiesis. Within 3 to 5 days new red blood cells are produced, increasing the hematocrit and the oxygen-carrying capacity. Thus, although the arterial P_{O_2} is not increased, the arterial oxygen content increases because of the increased blood hemoglobin concentration. This is at the cost of increased blood viscosity and ventricular work load. Increased concentrations of 2,3-DPG may help to release oxygen at the tissues.

Hypoxic stimulation of the arterial chemoreceptors persists indefinitely, although it may be somewhat diminished after prolonged periods at altitude. A more immediate finding is that the ventilatory response curve to carbon dioxide shifts to the left. That is, for any given alveolar or arterial P_{CO_2}, the ventilatory response is greater after several days at altitude. This increased response probably reflects alterations in central acid-base balance. It is associated with an alleviation of central nervous system symptoms and with a return of the cerebrospinal fluid pH toward normal because of a reduction of the bicarbonate concentration of the cerebrospinal fluid. Although it was initially believed that this bicarbonate reduction reflected active transport of the bicarbonate out of the cerebrospinal fluid, that belief is now somewhat controversial. Bicarbonate may instead simply diffuse out of the cerebrospinal fluid, or the reduced cerebrospinal fluid levels of bicarbonate may reflect decreased production of bicarbonate in the cerebrospinal fluid.

The elevated cardiac output, heart rate, and systemic blood pressure return to normal levels after (approximately) a month or so at altitude. This probably reflects a decrease in sympathetic activity or changes in sympathetic receptors. Nevertheless, hypoxic pulmonary vasoconstriction and pulmonary hypertension persist (along with increased blood viscosity), leading to right ventricular hypertrophy and frequently chronic cor pulmonale (right ventricular failure secondary to pulmonary hypertension).

DIVING AND THE RESPIRATORY SYSTEM

The major physiological stresses involved in diving include elevated ambient pressure, decreased effects of gravity, altered respiration, hypothermia, and sensory impairment. The severity of the stress involved depends on the depth attained, the length of the dive, and whether the breath is held or breathing apparatus is used. In this discussion we will concentrate on the first three physiological stresses.

Physical Principles

The pressure at the bottom of a column of liquid is proportional to the height of the column, the density of the liquid, and the acceleration of gravity. For example, for each 33 ft of seawater (or 34 ft of fresh water) ambient pressure increases by 1 atmosphere (atm). Thus, at a depth of 33 ft of seawater total ambient pressure is equal to 1520 torr.

The tissues of the body are mainly composed of water and are therefore nearly incompressible, but gases are compressible and follow Boyle's law. Thus, in a breath-hold dive the *volume* of gas in the lungs is inversely proportional to the depth attained. At 33 ft of depth (2 atm) lung volume is cut in half; at 66 ft (3 atm) it is one-third the original lung volume. As gases are compressed their *densities* increase.

As the total pressure increases, the *partial pressures* of the constituent gases also increase, according to Dalton's law. The biological effects of gases are generally dependent on their partial pressures rather than their fractional concentrations. Also, as the partial pressures of gases increase, the amounts *dissolved* in the tissues of the body increase, according to Henry's law.

Effects of Immersion up to the Neck

Merely immersing oneself up to the neck in water causes profound alterations in the cardiovascular and pulmonary systems. These effects are mainly a result of an increase in pressure outside of the thorax, abdomen, and limbs.

Mechanics of Breathing The pressure outside the chest wall of a person standing or seated in neck-deep water is greater than atmospheric, averaging about 20 cmH_2O. This positive pressure outside the chest opposes the normal outward elastic recoil of the chest wall and decreases the functional residual capacity. This occurs at the expense of the expiratory reserve volume, which may be decreased by as much as 70 percent. The intrapleural pressure is less negative at the functional residual capacity because of decreased outward elastic recoil of the chest wall. The work that must be done to bring air into the lung is greatly increased because extra inspiratory work is necessary to overcome the positive pressure outside the chest. Nonetheless, the vital capacity and total lung capacity are only slightly decreased. As was already pointed out, the expiratory reserve volume is decreased by neck-deep immersion, and the inspiratory reserve volume is therefore increased. The residual volume is slightly decreased because of an increase in pulmonary blood volume. Immersion up to the neck in water results in an increase in the work of breathing of about 60 percent.

Pulmonary Blood Flow During neck-deep immersion increased pressure outside of the limbs and the abdomen result in less pooling of systemic venous blood in gravity-dependent regions of the body. If the water temperature is below body temperature, a sympathetically mediated venoconstriction occurs, also augmenting venous return. The increased venous return increases the central blood volume by approximately 500 ml. Right atrial pressure increases from about -2 to $+16$ mmHg. As a result, the cardiac output and stroke volume increase by about 30 percent. The increase in pulmonary blood flow and pulmonary blood volume likely result in elevated mean pulmonary artery pressure, in capillary recruitment, in an increase in the diffusing capacity, and in a somewhat improved matching of ventilation and perfusion.

An additional effect of neck-deep immersion is the so-called immersion diuresis. Within a few minutes of immersion, urine flow increases four- to five-fold. Osmolal clearance increases only slightly. These findings are consistent with stimulation of stretch receptors in the left atrium and elsewhere in thoracic vessels by the increased thoracic blood volume. This, in turn, is believed to decrease the secretion of antidiuretic hormone (ADH) by the posterior pituitary gland.

Breath-hold Diving

During a breath-hold dive the total pressure of gases within the lungs is approximately equal to ambient pressure. Therefore the *volume* within the thorax must decrease proportionately and partial pressures of gases increase.

The Diving Reflex Many subjects demonstrate a profound bradycardia (decreased heart rate) and increased systemic vascular resistance with face immersion (especially into cold water). This "diving reflex" is initiated by as yet unknown sensors in the face or nose. A similar (but greater) response is seen when aquatic mammals such as whales and seals dive. The reflex decreases the work load of the heart and severely limits perfusion to all systemic vascular beds except for the strongest autoregulators, namely, the heart and brain. The cardiovascular effects of the diving reflex are similar to those produced by stimulation of the arterial chemoreceptors when no increase in ventilation can occur.

Gas Exchange in the Lungs Breath-hold divers usually hyperventilate before a dive so that typical alveolar P_{O_2}'s and P_{CO_2}'s might be 120 torr and 30 torr, respectively. Indeed, the breath-hold divers must take care not to hyperventilate so much that their P_{CO_2} gets so low that their "breakpoint" (mainly determined by Pa_{CO_2}) does not occur until after they lose consciousness from arterial hypoxemia. During a breath-hold dive to a depth of 33 ft lung volume decreases and gases are compressed. Total gas pressure approximately doubles: Thus after 20 s at 33 ft the alveolar P_{O_2} may be 160 to 180 torr; even after 1 min at 33 ft the alveolar PO_2 is well above 100 torr. The alveolar P_{CO_2}, however, also increases during descent to well above 40 torr and may increase even further. Thus, the transfer of oxygen from alveolus to blood is undisturbed until ascent; however, the normal transfer of carbon dioxide from blood to alveolus is reversed during descent and results in significant retention of carbon dioxide in the blood.

The Use of Underwater Breathing Apparatus

Self-Contained Underwater Breathing Apparatus, or SCUBA gear, mainly consists of a tank full of compressed gas that can be delivered by a demand regulator to the diver when the diver's mouth pressure decreases (during inspiration) to slightly less than the ambient pressure. Expired gas is simply released into the water as bubbles. Therefore, during a dive with scuba gear, gas pressure within the lungs remains close to the ambient pressure at a particular depth. The physiological stresses on the respiratory system of scuba diving are therefore mainly a consequence of elevated gas densities and partial pressures.

Mechanics of Breathing During scuba diving the inspiratory work of breathing is not a great problem at moderate depths because gas is delivered at ambient pressures. At very great depths, however, increased gas *density* becomes a problem because it elevates the airways' resistance work of breathing during turbulent flow. For example, in recent long-term experiments done with subjects *simulating* dives of over 2000 ft inside hyperbaric chambers, all subjects reported that they could breathe only through their mouths: The work of breathing through the nose was too great. This is one reason for replacing nitrogen with helium for deep dives. Helium is only about one-seventh as dense as nitrogen.

Control of Breathing The respiratory system sensitivity to carbon dioxide is decreased at great depths because of increased gas densities and high arterial P_{O_2}'s and because divers learn to suppress carbon dioxide drive to conserve compressed gas.

Other Hazards at Depth Other hazards that may be encountered in diving to great depths include barotrauma, decompression sickness, nitrogen narcosis, oxygen toxicity, and high-pressure nervous syndrome.

Barotrauma Barotrauma occurs when ambient pressure increases or decreases but the pressure in a closed unventilated area of the body that cannot equilibrate with ambient pressure does not. The barotrauma of descent is called "squeeze." It can affect the middle ear, if the eustachian tube is clogged or edematous, so that a person cannot equilibrate pressure in the middle ear; the sinuses; the lungs, resulting in pulmonary congestion, edema, or hemorrhage; and even cavities in the teeth. The barotrauma of ascent can occur if gases are trapped in areas of the body and begin to expand as the diver ascends. If a diver does not exhale as he or she ascends, expanding pulmonary gas may overdistend and rupture the lung ("burst lung"). This may result in hemorrhage, pneumothorax, or air embolism. Gases trapped in the gastrointestinal tract may cause abdominal discomfort and eructation or flatus as they expand. Barotrauma of the ears, sinuses, and teeth may also occur on rapid ascent from great depths.

Decompression Sickness Decompression sickness occurs when gas bubbles are formed in the blood and body tissues as the ambient pressure decreases. During a dive, the increasing ambient pressure causes an increase in the partial pressure of nitrogen in the body. The high partial pressure of nitrogen causes this normally poorly soluble gas to dissolve in the body tissues and fluids according to Henry's law. This is especially the case in body fat, which has a relatively high nitrogen solubility. At great depths, body tissues become supersaturated with nitrogen.

During a fast ascent, ambient pressure falls rapidly and nitrogen comes out of solution, forming bubbles in body tissues and fluids. The effect is the same as opening a bottle of a carbonated beverage such as a cola. During production of the soft drink it is exposed to higher than atmospheric pressures of gases, mainly carbon dioxide, and then capped. The total pressure in the gas layer above the liquid remains greater than atmospheric pressure. The partial pressures of gases dissolved in the liquid phase are in equilibrium with the partial pressures in the gas phase. Gases dissolve in the liquid phase according to Henry's law. When the bottle is uncapped, the pressure in the gas phase drops suddenly and the gas dissolved in the liquid phase comes out of solution, forming bubbles.

The bubbles formed in decompression sickness may form in the blood, in the joints of the extremities, and in the circulation of the central nervous system. Bubbles formed in the venous blood are usually trapped in the pulmonary circulation and rarely cause symptoms. The symptoms that do occasionally occur, which are known as "the chokes" by divers, include substernal chest pain, dyspnea, and cough and may be accompanied by pulmonary hypertension, pulmonary edema, and hypoxemia. These symptoms are obviously an extremely dangerous form of decompression sickness. Even more dangerous, of course, are bubbles in the circulation of the central nervous system, which may result in brain damage and paralysis. Osteonecrosis of joints may also be caused by inadequate decompression.

The treatment for decompression sickness is immediate recompression in a hyperbaric chamber, followed by slow decompression. Decompression sickness may be prevented by slow ascents from great depths (according to empirically derived decompression tables) and by substituting helium for nitrogen in inspired gas mixtures. Helium is only about one-half as soluble as nitrogen in body tissues.

Divers who ascend from submersion with no immediate effects of decompression may subsequently suffer decompression sickness if they ride in an airplane within a few hours of the dive. Commercial airplanes normally maintain cabin pressures well below 760 torr.

Nitrogen Narcosis Very high partial pressures of nitrogen directly affect the central nervous system, causing euphoria, loss of memory, clumsiness, and irrational behavior. This "rapture of the deep" occurs at depths of 100 ft or more and at greater depths may result in numbness of the limbs, disorientation, motor impairment, and ultimately unconsciousness. The mechanism of nitrogen narcosis is unknown.

Oxygen Toxicity Inhalation of 100% oxygen at 760 torr, or lower oxygen concentrations at higher ambient pressures, can cause both central nervous system and alveolar damage, although pulmonary manifestations are rare among divers. The mechanism of oxygen toxicity is controversial.

High-pressure Nervous Syndrome Exposure to very high ambient pressures, such as those encountered at very great depths (>250 ft), is associated with tremors and decreased manual dexterity. This high-pressure nervous syndrome (HPNS) usually occurs when nitrogen has been replaced by helium to decrease gas density, prevent nitrogen narcosis, and help avoid decompression sickness. Small amounts of nitrogen added to the inspired gas mixture help to counteract the problem. One hypothesis explaining HPNS is that the syndrome may result from a decreased lipid volume of nerves.

Study Questions

QUESTIONS

1 A subject inspires 500 ml from a spirometer. The subject's intrapleural pressure, determined using an esophageal balloon, was −5 cmH_2O before the inspiratory effort and was −10 cmH_2O at the end of the inspiration.

a What is the pulmonary compliance?

b What is the compliance of the chest wall?

c What is the total compliance?

$$\text{Compliance} = \frac{\Delta V}{P}$$

ANSWERS

a. Compliance of lungs is

$$C_L = \frac{\Delta V}{\underset{\text{(end inspiration)}}{(P_A - P_{ip})} - \underset{\text{(preinspiration)}}{(P_A - P_{ip})}}$$

where P_A = alveolar pressure
P_{ip} = intrapleural pressure

$$C_L = \frac{0.500 \text{ liter}}{[0 - (-10)cmH_2O] - [0 - (-5)\ cmH_2O]}$$

$$C_L = 0.10 \text{ liter}/cmH_2O$$

b. Compliance of chest wall is

Remember to use transmural pressure differences in calculations of compliance:

transmural pressure
= pressure inside
− pressure outside

$$Cw = \frac{\Delta V}{\underset{\text{(end inspiration)}}{(P_{ip} - P_{bs})} - \underset{\text{(preinspiration)}}{(P_{ip} - P_{bs})}}$$

where P_{ip} = intrapleural pressure
P_{bs} = body-surface pressure (atmospheric pressure in this problem)

$$Cw = \frac{0.500 \text{ liters}}{[-10 - (0)\ cmH_2O] - [-5 - (0)\ cmH_2O]}$$

$$Cw = -0.10 \text{ liter}/cmH_2O$$
(ignore minus sign)

c. Compliance of total respiratory system is

$$\frac{1}{C_T} = \frac{1}{C_L} + \frac{1}{Cw}$$

$$\frac{1}{C_T} = \frac{1}{0.10} + \frac{1}{0.10}$$

$$C_T = 0.05 \text{ liter}/cmH_2O$$

2 A postoperative patient whose respiratory muscles have been paralyzed with decamethonium, a curarelike drug, is maintained by a positive-pressure respirator. At end expiration (when alveolar pressure equals 0) intrapleural pressure, as measured by an esophageal balloon, is equal to $-3\ cmH_2O$. At the peak of inspiration alveolar pressure is $+20\ cmH_2O$ and intrapleural pressure is $+10\ cmH_2O$. Tidal volume is 500 ml.

a. What is the patient's pulmonary compliance?

b. What is the patient's thoracic compliance?

c. What is the patient's total compliance?

a. Since

$$C_L = \frac{0.500 \text{ liter}}{(20 - 10\ cmH_2O) - [0 - (-3)\ cmH_2O]}$$

$$C_L = 0.07 \text{ liter}/cmH_2O$$

b. Since

$$Cw = \frac{0.500 \text{ liter}}{(10 - 0\ cmH_2O) - (-3 - 0\ cmH_2O)}$$

$$Cw = 0.038 \text{ liter}/cmH_2O$$

c. Since

$$\frac{1}{C_T} = \frac{1}{0.07} + \frac{1}{0.038}$$

$$C_T = 0.025 \text{ liter}/cmH_2O$$

3 Which of the following conditions would be expected to *decrease* static pulmonary compliance (shift the pressure-volume curve for the lungs to the right)?

a Decreased functional pulmonary surfactant
b Fibrosis of the lungs
c Surgical removal of one lobe
d All of the above
e *a* and *b* only

d. All of the above lead to decreased compliance.

4 Which of the following conditions would be expected to *decrease* the functional residual capacity?

a Changing the body positions from standing to supine
b Pulmonary fibrosis
c Emphysema
d All of the above
e *a* and *b* only

e. *a* and *b* only—fibrosis increases lung elastic recoil, but emphysema decreases elastic recoil of the lungs. In the supine position the outward recoil of the chest wall is decreased.

5 Which of the following tend to *increase* airways resistance?

a Stimulation of the parasympathetic postganglionic fibers innervating the bronchial and bronchiolar smooth muscle
b Low lung volumes
c Forced expirations
d Breathing through the nose instead of the mouth
e All of the above

e. All of the above

6 What is the effect on each of the following standard lung volumes and capacities of changing from the supine position to standing upright?

FRC (functional residual capacity)
RV (residual volume)
ERV (expiratory reserve volume)

As a person stands up, the effects of gravity play a much greater role in the mechanics of breathing (and also decrease venous return). The contents of the abdomen are pulled away from the diaphragm, thus increasing the outward elastic recoil of the chest wall. The inward recoil of the lungs is not affected, and so

TLC (total lung capacity)
V_T (tidal volume)
IRV (inspiratory reserve volume)
IC (inspiratory capacity)
VC (vital capacity)

the *functional residual capacity* is increased. The *residual volume* is relatively unaffected. The *expiratory reserve volume* increases because the FRC is increased and the residual volume is relatively unchanged. The *total lung capacity* may increase *slightly* because of the slightly decreased inward elastic recoil of the chest wall at high lung volumes and because the abdominal contents are pulled away from the diaphragm. The *tidal volume* is probably unchanged or possibly slightly increased again because of the increased outward elastic recoil of the chest wall at the FRC. The higher FRC and similar TLC and V_T lead to a decrease in the *inspiratory reserve volume* and a decrease in the *inspiratory capacity*. The *vital capacity* is also relatively unchanged, although it may be slightly increased because of the slight increase in TLC and decreased intrathoracic blood volume.

7 Predict the effects of old age on the standard lung volumes and capacities and the closing volume.

Assuming general health and normal weight, the main changes seen with age are a *loss of pulmonary elastic recoil* and a slight *increase* of the elastic recoil of the chest wall, especially at higher volumes. The loss of pulmonary elastic recoil has the secondary effect of *increasing airway closure* in dependent areas of the lung at the lower lung volumes. For these reasons the *FRC* will be increased and the *RV* may be greatly increased, with the *TLC* slightly decreased. The V_T should be unchanged or may either be slightly increased or decreased, depending on whether the increased lung compliance, increased airways resistance, or decreased chest wall compliance predominates. The *ERV* will decrease because the increase in *RV* due to airway closure is greater than the increase in FRC. *IRV* and *IC* are decreased as is the *VC*. The *closing volume* is also increased.

8 A volume of 1 liter of gas is measured in a spirometer at 23°C (296K) (P_{H_2O} is 21 mmHg) and the barometric pressure is 770 mmHg.

a What would the volume be under STPD conditions?

b What would the volume be under BTPS conditions?

$V_2 = V_1 \times$ temperature correction $\times$ pressure correction

Temperature correction =

$$\frac{V_2}{V_1} = \frac{T_2}{T_1}$$

Therefore,

$$V_2 = V_1 \times \frac{T_2}{T_1}$$

Pressure correction = $P_1V_1 = P_2V_2$

Therefore,

$$V_2 = V_1 \times \frac{P_1}{P_2}$$

Combining the two, we get

$$V_2 = V_1 \times \frac{T_2}{T_1} \times \frac{P_1}{P_2}$$

a. The V_{STPD} is

$$V_{STPD} = V_{ATPS} \times \frac{T_{STPD}}{T_{ATPS}} \times \frac{P_{ATPS}}{P_{STPD}}$$

$$V_{STPD} = 1 \text{ liter} \times \frac{273 \text{ K}}{(273 + 23) \text{ K}} \times \frac{(770 - 21) \text{ torr}}{760 \text{ torr}}$$

$$V_{STPD} = 0.91 \text{ liter}$$

b. The V_{BTPS} is

$$V_{BTPS} = V_{ATPS} \times \frac{T_{BTPS}}{T_{ATPS}} \times \frac{P_{ATPS}}{P_{BTPS}}$$

$$V_{BTPS} = 1 \text{ liter} \times \frac{(273 + 37)\text{ K}}{(273 + 23)\text{ K}} \times \frac{(770 - 21)\text{ torr}}{(770 - 47)\text{ torr}}$$

$$V_{BTPS} = 1.1 \text{ liter}$$

9 A subject starts at her FRC and breathes 100% O_2 through a one-way valve. Her expired air is collected in a very large spirometer (called a *Tissot spirometer*). The test is continued until her expired N_2 concentration, as measured by a nitrogen analyzer, is virtually zero. At this time there are 36 liters of gas in the spirometer, of which 5.6% is N_2. What is the subject's FRC?

The volume of N_2 in the spirometer is 0.0556 × 36 liters, or 2.0 liters. This is the volume of N_2 in the subject's lungs when the test began (at her FRC). Since N_2 constituted 80% of her FRC, then her FRC is equal to 100/80 × 2.0 liters, or 1.25 × 2.0 liters, which is equal to 2.5 liters.

10 A subject rebreathes the gas in a 20-liter capacity spirometer that originally contained 10 liters of 15% helium. After a number of breaths the concentration of helium in the subject's lungs is equal to that now in the spirometer, which is 11% helium. (During the equilibration period the expired CO_2 was absorbed by an absorbent chemical in the spirometer and O_2 was added to the spirometer at the subject's $\dot{V}O_2$.) *At the end of a normal expiration* the spirometer contains 10.64 liters when corrected to BTPS. What is the subject's FRC?

Since helium is not absorbed or given off by the lung, the initial amount of helium in the system must equal the final amount of helium in the system. The amount is equal to the fractional concentration × the volume:

$$F_{HE_i} \times Vsp_i = F_{HE_f}(Vsp_f + V_{L_f})$$

where F = fractional concentration
V = volume
sp = spirometer
L = lungs
i = initial

f = final (after equilibration)

Because the test was ended at the end of a normal expiration, VL_f equals the subject's FRC.

$$0.15 \times 10 \text{ liters} = 0.11\,(10.64 \text{ liters} + \text{FRC})$$

$$0.15 \text{ liters} = 1.17 \text{ liters} + 0.11 \times \text{FRC}$$

$$0.11 \times \text{FRC} = 1.5 \text{ liters} - 1.17 \text{ liters}$$

$$\text{FRC} = \frac{0.33 \text{ liter}}{0.11}$$

$$\text{FRC} = 3.0 \text{ liters}$$

11 A subject in a body plethysmograph breathes normally through a mouthpiece. At the end of a normal expiration a valve in the mouthpiece is closed. The next inspiratory effort is made against the closed valve. Additional air cannot enter the lungs; instead, the inspiratory effort lowers the pressure at the mouth by 10 torr and expands the gas in the lungs by 50 ml, as determined by the increase in the plethysmograph pressure and its calibration curve with the subject in the box. What is the subject's FRC?

$$PM_i \times VL_i = PM_f \times (VL_i + \Delta V)$$

where P = pressure
V = volume
M = mouth
L = lung
i = initial
f = final

Because the valve was closed at the end of a normal expiration, VL_i equals the subject's FRC.

$$760 \text{ torr} \times \text{FRC} = 750 \text{ torr}\,(\text{FRC} + 50 \text{ ml})$$

$$10 \text{ FRC} = 37{,}500 \text{ ml}$$

$$\text{FRC} = 3.75 \text{ liters}$$

12 A subject breathes at a rate of 10 breaths per minute. His tidal volume is 500 ml.

a What is $\dot{V}E$?

b If his anatomic dead space

a. $\dot{V}E$ is

$$\dot{V}E = n \times VT$$

$$= \frac{10 \text{ breaths}}{\text{min}} \times \frac{500 \text{ ml}}{\text{breath}}$$

was determined to be 150 ml using Fowler's method, what is his $\dot{V}_A$?

c If he increases his rate to 15 breaths per minute with V_T remaining at 500 ml, what will his new $\dot{V}_E$ and $\dot{V}_A$ be?

d If he increases his V_T to 750 ml, with his rate remaining at 10 breaths per minute, what will his new $\dot{V}_E$ and $\dot{V}_A$ be?

$$= \frac{5000 \text{ ml}}{\text{min}}$$

b. $\dot{V}_A$ is

$$V_T = V_A + V_D$$

Multiply by rate

$$\dot{V}_E = \dot{V}_A + \dot{V}_D$$

$$\dot{V}_A = \dot{V}_E - \dot{V}_D = \frac{10 \text{ breaths}}{\text{min}}\left(\frac{500 \text{ ml}}{\text{breath}} - \frac{150 \text{ ml}}{\text{breath}}\right)$$

$$\dot{V}_A = \frac{3500 \text{ ml}}{\text{min}}$$

c. The new $\dot{V}_E$ and $\dot{V}_A$ are

$$V_E = \frac{15 \text{ breaths}}{\text{min}} \times \frac{500 \text{ ml}}{\text{breath}} = \frac{7500 \text{ ml}}{\text{min}}$$

$$\dot{V}_A = \frac{15 \text{ breaths}}{\text{min}}\left(\frac{500 \text{ ml}}{\text{breath}} - \frac{150 \text{ ml}}{\text{breath}}\right) = \frac{5250 \text{ ml}}{\text{min}}$$

d. The new $\dot{V}_E$ and $\dot{V}_A$ are

$$V_E = \frac{10 \text{ breaths}}{\text{min}} \times \frac{750 \text{ ml}}{\text{breath}} = \frac{7500 \text{ ml}}{\text{min}}$$

$$\dot{V}_A = \frac{10 \text{ breaths}}{\text{min}}\left(\frac{750 \text{ ml}}{\text{breath}} - \frac{150 \text{ ml}}{\text{breath}}\right) = \frac{6000 \text{ ml}}{\text{min}}$$

13 The following measurements were made on a patient when the barometric pressure was 747 torr:

$$\frac{V_D}{V_T} = \frac{Pa_{CO_2} - PE_{CO_2}}{Pa_{CO_2}}$$

$$PE_{CO_2} = 0.04\,(747 \text{ torr} - 47 \text{ torr})$$

$$Pa_{CO_2} = 40 \text{ torr}$$
$$FE_{CO_2} = 0.04$$
$$\dot{V}E = \frac{6 \text{ liters}}{\text{min}}$$
$$\text{breathing frequency} = \frac{12 \text{ breaths}}{\text{min}}$$

where FE_{CO_2} = fractional concentration of CO_2 in the subject's mixed expired air

What is the patient's VD/VT? What is the patient's physiological dead space?

$$PE_{CO_2} = 28 \text{ torr}$$
$$\frac{VD}{VT} = \frac{40 \text{ torr} - 28 \text{ torr}}{40 \text{ torr}}$$
$$\frac{VD}{VT} = 0.3$$

Since

$$VT = \frac{\dot{V}E}{n}$$
$$VT = \frac{6000 \text{ ml/min}}{12 \text{ breaths/min}}$$
$$VT = 500 \text{ ml}$$
$$VD = 0.3 \times 500 \text{ ml}$$
$$VD = 150 \text{ ml}$$

14 A person with a PA_{CO_2} of 40 torr, a PA_{O_2} of 104 torr, and a respiratory exchange ratio of 0.8 breathing room air at a barometric pressure of 760 torr doubles alveolar ventilation. What will this person's new steady-state PA_{CO_2} and PA_{O_2} be (assuming no change in oxygen consumption and carbon dioxide production)?

$$PA_{CO_2} \ \alpha \ \frac{\dot{V}CO_2}{\dot{V}A}$$

If alveolar ventilation doubles, alveolar PCO_2 is cut in half:

$$PA_{CO_2} = 20 \text{ torr}$$
$$PA_{O_2} = PI_{O_2} - \frac{PA_{CO_2}}{R} + F$$
$$PI_{O_2} = 0.2093 \times (760 - 47) \text{ torr}$$
$$PI_{O_2} = 149 \text{ torr}$$
$$PA_{O_2} = 149 \text{ torr} - \frac{20 \text{ torr}}{0.8} + F$$
$$PA_{O_2} = 149 \text{ torr} - 25 \text{ torr} = 124 \text{ torr}$$

15 A normal person, seated upright, begins to inspire from *residual volume*. The first 100 ml of inspired gas is labeled with xenon 133. Most of this radioactive gas (i.e., the first 100 ml of gas inspired after the dead space) would likely be found:

a In alveoli in lower portions of the lung

Answer is *b*. At the residual volume airways in gravity-dependent portions of the lungs are likely to be collapsed. Alveoli in upper regions of the lung are on the steep portion of their pressure-volume curves (i.e., they are more compliant than they are at higher lung volumes), and so most of the labeled gas will

b In alveoli in upper portion of the lung
c Uniformly distributed to all alveoli

enter alveoli in the upper portions of the lung.

16 Using the data provided in Fig. 4-1 and assuming a cardiac output of 5 liters/min, calculate the pulmonary vascular resistance and the systemic vascular resistance.

$$\Delta P = \dot{Q} \times R$$

$$R = \frac{\Delta P}{\dot{Q}}$$

Pulmonary vascular resistance (PVR):

$$PVR = \frac{MPAP - MLAP}{CO}$$

where MPAP = mean pulmonary artery pressure
MLAP = mean left atrial pressure
CO = cardiac output

$$PVR = \frac{(15 - 5) \text{ mmHg}}{5 \text{ liters/min}}$$

$$PVR = 2 \text{ mmHg/liter/min}$$

Systemic vascular resistance (SVR):

$$SVR = \frac{MABP - RAP}{CO}$$

where MABP = mean arterial blood pressure
RAP = right atrial pressure
CO = cardiac output

$$SVR = \frac{(100 - 2) \text{ mmHg}}{5 \text{ liters/min}}$$

$$SVR = 19.6 \text{ mmHg/liter/min}$$

17 Which of the following situations would be expected to *decrease* pulmonary vascular resistance?

a Ascent to 15,000 ft above sea level

b Inspiration to the total lung capacity

c Expiration to the residual volume

d Moderate exercise

e Blood loss secondary to trauma

Only moderate exercise (d) would be expected to cause a decrease in pulmonary vascular resistance. Exercise increases the cardiac output and raises pulmonary artery pressure, which cause recruitment of pulmonary vessels, distention of pulmonary vessels, or both.

At 15,000 feet above sea level, barometric pressure is only about 429 torr (mmHg). Breathing ambient air, the inspired P_{O_2} is about 80 torr, and alveolar P_{O_2} is about 50 torr. This degree of alveolar hypoxia elicits hypoxic pulmonary vasoconstriction and increases the pulmonary vascular resistance.

Pulmonary vascular resistance is high at both high and low lung volumes, as shown in Fig. 4-3.

Blood loss leads to decreased venous return and a decrease in cardiac output and mean pulmonary artery pressure. This causes a derecruitment of pulmonary vessels and decreases their distention, increasing pulmonary vascular resistance.

18 Which of the following situations would be expected to lead to an *increase* in the amount of the lung under zone 1 conditions?

a Ascent to 15,000 ft above sea level

b Blood loss secondary to trauma

c Moderate exercise

d Positive-pressure ventilation with positive end-expiratory pressure (PEEP)

e Changing from the standing to the supine position

Zone 1 is defined as an area of the lung in which no blood flow occurs because alveolar pressure is greater than pulmonary artery pressure. Blood loss secondary to trauma lowers venous return and cardiac output. As a result, pulmonary artery pressure is likely to fall, increasing the likelihood of zone 1 conditions in the lung. Positive-pressure ventilation with PEEP (positive airway pressures during *expiration,* often used on patients prone to spontaneous atelectasis, leads to positive alveolar and pleural pressures throughout the respiratory cycle) would also increase the likelihood of zone 1 conditions in the lung.

Ascent to 15,000 feet increases pulmonary artery pressure by activating hypoxic pulmonary vasoconstriction; moderate exercise increases mean pulmonary artery pressure by increasing venous return. Both of these would decrease the tendency toward zone 1 conditions. Lying down lowers the hydrostatic pressure gradient that must be overcome to perfuse nondependent portions of the lung. It also increases, at least transiently, venous return and cardiac output by decreasing the amount of blood held in the systemic veins by gravity.

19 Which of the following circumstances might be expected to lead to pulmonary edema?

a Overtransfusion with saline

b Occlusion of the lymphatic drainage of an area of the lung

c Left ventricular failure

d Low concentration of plasma proteins

e Destruction of portions of the pulmonary capillary endothelium by toxins

Each of the above conditions could contribute to the formation of pulmonary edema. Left ventricular failure and overtransfusion with saline both increase pulmonary capillary hydrostatic pressure, which increases the tendency toward pulmonary edema, as given by the Starling equation. Low plasma protein concentration, caused by a protein-poor diet or renal problems, or by *dilution* in overtransfusion with saline, is another predisposing factor that may lead to pulmonary edema because it lowers the plasma colloid osmotic pressure. Destruction of portions of the pulmonary capillary endothelium or occlusion of the lymphatic drainage of portions of the lung may also be causative factors of pulmonary edema.

20 An otherwise normal person accidentally aspirates a foreign body into the right main stem bronchus, partially occluding it. Which of the following is/are likely to occur?

a The right lung $P_{A_{O_2}}$ will be lower and $P_{A_{CO_2}}$ will be higher than those of the left lung.

b The calculated shunt fraction, $\dot{Q}s/\dot{Q}_T$, will increase.

c Blood flow to the right lung will decrease.

d The arterial P_{O_2} will fall.

e All of the above are correct.

All of the above answers are correct. With a partial obstruction of its airway the right lung will have a lower ventilation-perfusion ratio than that of the left lung; therefore, it will have a lower alveolar P_{O_2} and a higher alveolar P_{CO_2}. With more alveolar-capillary units overperfused, the calculated shunt will increase. The overperfusion may be somewhat attenuated if *hypoxic pulmonary vasoconstriction* diverts some blood flow away from hypoxic and hypercapnic alveoli to the better ventilated left lung, but this response never functions perfectly. As a result the arterial P_{O_2} will fall.

21 A normal person lies down *on her right side* and breathes normally. Her right lung, in comparison to her left lung, will be expected to have a:

a Lower $P_{A_{O_2}}$ and a higher $P_{A_{CO_2}}$

b Higher blood flow per unit volume

c Greater ventilation per unit volume

d Higher ventilation-per-fusion ratio

e Larger alveoli

a, b, and *c* above are correct; d and e are incorrect. The right lung, which is more gravity-dependent, will have a greater blood flow per unit volume than will the left lung because hydrostatic forces increase the intravascular pressures, causing more distention, recruitment, or both. The pleural surface pressure is less negative in the more gravity-dependent region, and so the alveolar-distending pressure is less in the right lung and the alveoli are smaller. Because of this, the alveoli of the right lung are on a steeper portion of their pressure-volume curves and are therefore better ventilated. The difference in blood flow between the two lungs, however, is greater than is the difference in ventilation, and so the right lung has a lower $\dot{V}_A/\dot{Q}c$ than does the left. This leads to a lower $P_{A_{O_2}}$ and a higher $P_{A_{CO_2}}$ in the right lung.

22 How would each of the following conditions or circumstances be expected to affect the diffusing capacity (D_L) of the lungs? Explain your answers.

a Changing from the supine to the upright position

b Exercise

c Valsalva maneuver

d Anemia

e Low cardiac output due to blood loss

f Diffuse interstitial fibrosis of the lungs

g Emphysema

a. Changing from the supine to the upright position slightly decreases the diffusing capacity by decreasing the venous return, because of pooling of blood in the extremities and abdomen. The decreased venous return decreases the central blood volume and may slightly decrease the right ventricular output, resulting in derecruitment of pulmonary capillaries and decreased surface area for diffusion.

b. Exercise increases the diffusing capacity by increasing the cardiac output. This recruits previously unperfused capillaries, increasing the surface area available for diffusion. Oxygen transfer across the alveolar-capillary barrier will also increase because at high cardiac outputs the linear velocity of the blood moving through the pulmonary capillaries increases and there is less perfusion limitation of oxygen transfer.

c. A Valsalva maneuver (an expiratory effort against a closed glottis) greatly decreases the pulmonary capillary blood volume and therefore decreases the diffusing capacity.

d. Anemia decreases the diffusing capacity by decreasing the hemoglobin available to chemically combine with oxygen. The partial pressure of oxygen in the plasma in the pulmonary capillaries therefore equilibrates more rapidly with the alveolar P_{O_2}, leading to increased perfusion limitation of oxygen transfer.

e. A low cardiac output due to blood loss decreases the diffusing capacity by decreasing the venous return and the central blood volume. Pulmonary capillary blood volume decreases, resulting in derecruitment and decreased surface area for diffusion.

f. Diffuse interstitial fibrosis of the lungs thickens the alveolar-capillary barrier (as does interstitial edema), resulting in decreased diffusion of gases across the alveolar-capillary barrier, in accordance with Fick's law.

g. Emphysema destroys the alveolar interstitium and blood vessels, decreasing the surface area for diffusion.

23 An otherwise normal person has lost enough blood to decrease his hemoglobin concentration from 15 g per 100 ml blood to 12 g per 100 ml blood. Which of the following is expected to decrease?

a Arterial P_{O_2}

b Blood oxygen-carrying *capacity*

c Arterial hemoglobin *saturation*

d Arterial oxygen *content*

b and *d* only. The blood oxygen-carrying *capacity* (excluding physically dissolved O_2) will decrease from 20.1 to 16.08 ml O_2 per 100 ml blood. The arterial PO_2 is not affected by the decreased hemoglobin concentration, and so the oxygen *saturation* of the hemoglobin that *is* present is also unaffected. If alveolar Po_2 is approximately 104 torr, then the arterial Po_2 is about 100 and hemoglobin is about 97.4 percent saturated. The arterial oxygen *content* (excluding physically dissolved O_2) is therefore reduced from 19.58 to 15.66 ml O_2 per 100 ml blood.

24 Results of tests on a patient's blood indicate her hemoglobin concentration to be 10 g per 100 ml blood. Her blood is 97.4 percent saturated with oxygen at a Pa_{O_2} of 100 torr. What is her arterial oxygen content, including physically

Physically dissolved:

$$\frac{0.003 \text{ ml } O_2/100 \text{ ml blood}}{\text{torr } Po_2} \times 100 \text{ torr} = \frac{0.3 \text{ ml } O_2}{100 \text{ ml blood}}$$

dissolved oxygen (at 37°C, pH of 7.40, P_{CO_2} of 40 torr)?

Bound to hemoglobin:

$$\frac{1.34 \text{ ml } O_2}{\text{gram Hb}} \times \frac{10 \text{ g Hb}}{100 \text{ ml blood}} \times 0.974 = \frac{\textit{13.05 ml } O_2}{\textit{100 ml blood}}$$

Total: 13.35 ml O_2/100 ml blood

25 What is the approximate hemoglobin oxygen saturation (S_{O_2}) of a blood sample that contains 10 g Hb per 100 ml blood and oxygen content of 10 ml O_2 per 100 ml blood (ignore physically dissolved O_2)?

Oxygen-carrying *capacity:*

$$\frac{1.34 \text{ ml } O_2}{\text{gram Hb}} \times \frac{10 \text{ g Hb}}{100 \text{ ml blood}} = \frac{13.4 \text{ ml } O_2}{100 \text{ ml blood}}$$

$$\text{saturation} = \frac{\text{content}}{\text{capacity}} = \frac{10}{13.4} = 0.75 \times 100\%$$

saturation = 75%

For each of the following sets of blood gas data, calculate the plasma bicarbonate concentration and then attempt to identify the underlying disorder. Assume the body temperature to be 37°C and the hemoglobin concentration to be 15 g of Hb per 100 ml blood. $F_{I_{O_2}}$ is 0.21 (room air). Use Figs. 8-1, 8-3, and 8-5 to assist you in making your "diagnosis."

26 Pa_{O_2} = 89 torr; Pa_{CO_2} = 40 torr; and pHa = 7.20

$$\text{pH} = 6.1 + \log \frac{[HCO_3^-]p}{0.03 \times Pa_{CO_2}}$$

$$7.2 = 6.1 + \log \frac{[HCO_3^-]p}{1.2}$$

$$1.1 = \log \frac{[HCO_3^-]p}{1.2}$$

$$1.1 = \log [HCO_3^-]p - \log 1.2$$

$$1.1 = [HCO_3^-] - 0.08$$

$$1.18 = \log [HCO_3^-]p$$

$$15.11 \text{ meq/liter} = [HCO_3^-]p$$

The pHa is below normal limits; the Pa_{CO_2} is normal; and the plasma bicarbonate concentration is low. The Pa_{O_2} is not remarkable, especially if this is an older patient. (No one would make a diagnosis on blood gas data alone.) This is probably uncompensated metabolic acidosis.

27 $Pa_{O_2} = 50$ torr; $Pa_{CO_2} = 60$ torr; and pHa = 7.27

$[HCO_3^-]p = 26.9$ meq/liter

The pHa is low; the Pa_{CO_2} is high; and the plasma bicarbonate concentration is slightly elevated. The Pa_{O_2} is low. This appears to be uncompensated respiratory acidosis secondary to hypoventilation, caused for example by acute respiratory depression or acute airways obstruction.

28 $Pa_{O_2} = 95$ torr; $Pa_{CO_2} = 38$ torr; and pHa = 7.52

$[HCO_3^-]p = 30.2$ meq/liter

The pHa is elevated and the Pa_{CO_2} and Pa_{O_2} are within normal limits. Bicarbonate is elevated. This appears to be uncompensated metabolic alkalosis.

29 $Pa_{O_2} = 120$ torr; $Pa_{CO_2} = 20$ torr; and pHa = 7.62

$[HCO_3^-]p = 19.9$ meq/liter

The pHa is very high and Pa_{CO_2} is very low. Pa_{O_2} is abnormally *high*. The plasma bicarbonate concentration is slightly depressed, but it falls on the normal buffer line for a pH of 7.6. This is uncompensated respiratory alkalosis secondary to hyperventilation. Since the Pa_{O_2} is *high,* it is not caused by hypoxic stimulation of alveolar ventilation (see Chap. 11) but is likely of voluntary or psychological origin or caused by drugs or overadministration of mechanical ventilation.

30 $Pa_{O_2} = 45$ torr; $Pa_{CO_2} = 60$ torr; and pHa = 7.36

$[HCO_3^-]p = 33.1$ meq/liter

The arterial pH is low but within normal limits. The arterial P_{O_2} is very low and the arterial P_{CO_2} is very high. The plasma bicarbonate concentration is above the normal buffer line. This is chronic respiratory acidosis (caused by hypoventilation) with renal compensation. This pattern is a familiar one in chronic obstructive lung disease.

31 $Pa_{O_2} = 105$ torr; $Pa_{CO_2} = 30$ torr; and pHa = 7.34

$[HCO_3^-]p = 15.49$ meq/liter

The arterial pH is low, indicating acidosis, but arterial PCO_2 is also low. The arterial PO_2 is high and plasma bicarbonate concentration is low. This appears to be metabolic acidosis with respiratory compensation, as indicated by the elevated Pa_{O_2} and decreased Pa_{CO_2}.

32 $Pa_{O_2} = 62$ torr; $Pa_{CO_2} = 50$ torr; and pHa = 7.22

$[HCO_3^-]p = 19.8$ meq/liter

The pHa is low and Pa_{CO_2} is elevated; nonetheless, the plasma bicarbonate concentration is *also* depressed. Arterial PO_2 is low. This appears to be *mixed* respiratory and metabolic acidosis.

33 Voluntary apnea for 90 s will:

a Increase arterial Pco_2

b Decrease arterial Po_2

c Stimulate the arterial chemoreceptors

d Stimulate the central chemoreceptors

All of the above are correct. Voluntary apnea (breath-holding) for 90 s causes alveolar Po_2 to decrease and alveolar Pco_2 to increase (about 15 torr). These alterations are reflected in the arterial Po_2 and Pco_2. The decrease in arterial Po_2 and the increase in arterial Pco_2 (and increase in hydrogen ion concentration) stimulate the arterial chemoreceptors. Ninety seconds is a sufficient period for the carbon dioxide to begin to diffuse into the cerebrospinal fluid and stimulate the central chemoreceptors. The hypoxia and acidosis should have little effect on the central chemoreceptors: The central chemoreceptors are not responsive to hypoxia and few hydrogen ions would be expected to get across the blood-brain barrier in 90 seconds.

34 Which of the following conditions would be expected to increase alveolar ventilation by stimulating the arterial chemoreceptors?

a Mild anemia

b Severe exercise

c Hypoxia due to ascent to high altitude

d Acute airway obstruction

e Large intrapulmonary shunts

All but *a* above are correct. Mild anemia without metabolic acidosis does not stimulate the arterial chemoreceptors because the arterial chemoreceptors are stimulated by low arterial Po_2 rather than low arterial oxygen content. Severe exercise may cause a lactic acidosis that stimulates the arterial chemoreceptors. Intrapulmonary shunts and hypoxia stimulate the arterial chemoreceptors, as does acute airway obstruction, which leads to hypoxia and hypercapnia.

35 Which of the following would be expected to occur as an untrained person begins to exercise?

a Decreased pulmonary vascular resistance

All of the above are expected findings in exercise. As cardiac output increases in response to exercise, previously unperfused capillaries are recruited. This recruitment,

b Increased cardiac output
c More homogeneous ventilation-perfusion ratios through the lung
d Increased diffusing capacity

which occurs primarily in upper regions of the lung, results in a decreased pulmonary vascular resistance, more homogeneous $\dot{V}_A/\dot{Q}$'s, and a greater surface area for diffusion. The increased cardiac output also results in increased linear velocity of blood flow through the lung, which helps to increase the diffusing capacity.

36 Which of the following responses would be expected by a normal person after 6 days of residence at an altitude of 15,000 ft?
a Elevated mean pulmonary artery pressure
b Alveolar ventilation greater than that at sea level
c Increased hematocrit
d Decreased plasma bicarbonate
e Normal arterial P_{CO_2}

All but *e* above are correct. Mean pulmonary artery pressure is elevated because of the continued presence of hypoxic pulmonary vasoconstriction. Alveolar ventilation continues to be elevated because of hypoxic drive of the arterial chemoreceptors. This results in hypocapnia. The respiratory alkalosis caused by the hypocapnia, however, is partially compensated for by renal excretion of base, resulting in a decreased plasma bicarbonate concentration. Hematocrit is increased secondary to increased erythropoiesis as mediated by erythropoietin.

37 Which of the following are expected consequences of neck-deep immersion in water?
a Increased work of breathing
b Decreased functional residual capacity
c Increased expiratory reserve volume
d Decreased inspiratory reserve volume

a and *b* above are correct but c and d above are incorrect. The outward elastic recoil of the chest wall is decreased by the hydrostatic pressure of neck-deep water. This increases the work that must be done to bring air into the lungs and decreases the functional residual capacity. The decrease in functional residual capacity occurs mainly at the expense of the expiratory reserve volume, which decreases. Because the vital capacity decreases only slightly, the inspiratory reserve volume is increased.

Suggested Readings

CHAPTER 1

Bloom, W., and D. W. Fawcett: *A Textbook of Histology,* 9th ed., Saunders, Philadelphia, 1968, pp. 621–659.

Bouhuys, A.: *The Physiology of Breathing,* Grune & Stratton, New York, 1977, pp. 26–42.

Gehr, P., M. Bachofen, and E. R. Weibel: "The Normal Human Lung: Ultrastructure and Morphometric Estimation of Diffusion Capacity," *Respir. Physiol.,* **32:**121 (1978).

Krahl, V. E.: "Anatomy of the Human Lung," in W. O. Fenn and H. Rahn (eds.), *Handbook of Physiology,* vol. 1, sec. 3: "Respiration," American Physiological Society, Washington, D.C., 1964, pp. 213–284.

Netter, F. H.: *The Ciba Collection of Medical Illustrations,* vol. 7: *Respiratory System,* Ciba, Summit, N.J., 1979, pp. 3–43.

Proctor, D. F.: "The Upper Airways," *Am. Rev. Respir. Dis.,* **115:**97, 315 (1977).

Weibel, E. R.: *Morphometry of the Human Lung,* Springer-Verlag, Berlin, 1963.

______: "Morphometric Estimation of Pulmonary Diffusing Capacity, I. Model and Method," *Respir. Physiol.,* **11:**54 (1970).

———: "Morphological Basis of Alveolar-Capillary Gas Exchange," *Physiol. Rev.*, **53**:419 (1973).

———: "Design and Structure of the Human Lung," in A. P. Fishman (ed.), *Pulmonary Diseases and Disorders,* vol. 1, McGraw-Hill, New York, 1980, pp. 224–271.

CHAPTER 2

Agostoni, E.: "Mechanics of the Pleural Space," *Physiol. Rev.*, **52**:57–128 (1972).

———, and J. Mead: "Statics of the Respiratory System," in W. O. Fenn and H. Rahn (eds.), *Handbook of Physiology,* vol. 1, sec. 3: "Respiration," American Physiological Society, Washington, D.C., 1964, pp. 389–407.

Altose, M. D.: "The Physiological Basis of Pulmonary Function Testing," *Ciba Clinical Symposia,* **31**:1–39 (1979).

———: "Pulmonary Mechanics," in A. P. Fishman (ed.), *Pulmonary Diseases and Disorders,* vol. 1, McGraw-Hill, New York, 1980, pp. 359–372.

Bouhuys, A.: *The Physiology of Breathing,* Grune & Stratton, New York, 1977, pp. 60–79, 173–232.

Clements, J. A., and D. F. Tierney: "Alveolar Instability Associated with Altered Surface Tension," in W. O. Fenn and H. Rahn (eds.), *Handbook of Physiology,* vol. 2, sec. 3: "Respiration," American Physiological Society, Washington, D.C., 1965, pp. 1567–1568.

Comroe, J. H.: *Physiology of Respiration,* 2d ed., Year Book, Chicago, 1974, pp. 94–141.

Green, J. F.: *Mechanical Concepts in Cardiovascular and Pulmonary Physiology,* Lea & Febiger, Philadelphia, 1977, pp. 81–113.

Hyatt, R. E.: "Dynamic Lung Volumes," in W. O. Fenn and H. Rahn (eds.), *Handbook of Physiology,* vol. 2, sec. 3: "Respiration," American Physiological Society, Washington, D.C., 1965, pp. 1381–1398.

Macklem, P. T., and B. R. Murphy: "The Forces Applied to the Lung in Health and Disease," *Am. J. Med.,* **57**:371–377 (1974).

Mead, J., T. Takashima, and D. Keith: "Stress Distribution in Lungs: A Model of Pulmonary Elasticity," *J. Physiol.,* **28**:596–608 (1970).

———, J. M. Turner, P. T. Macklem, and J. B. Little: "Significance of the Relationship Between Lung Recoil and Maximum Expiratory Flow," *J. Appl. Physiol.,* **22**:95–108 (1967).

Murray, J. F.: *The Normal Lung,* Saunders, Philadelphia, 1976, pp. 77–107.

———, R. H. Greenspan, W. M. Gold, and A. B. Cohen: "Early Diagnosis of Chronic Obstructive Lung Disease," *California Medicine,* **116**:37–55 (1972).

Nunn, J. F. *Applied Respiratory Physiology,* 2d ed., Butterworths, London, 1977, pp. 63–93, 238–240.

Otis, A. B.: The Work of Breathing," in W. O. Fenn and H. Rahn (eds.), *Handbook of Physiology,* vol. 1, sec. 3: "Respiration," American Physiological Society, Washington, D.C., 1964, pp. 463–475.

Radford, E. P.: *Tissue Elasticity,* American Physiological Society, Washington, D.C., 1957.

CHAPTER 3

Agostoni, E.: "Mechanics of the Pleural Space," *Physiol. Rev.,* **52:**57–128 (1972).

Altose, M. D.: "The Physiological Basis of Pulmonary Function Testing," *Ciba Clinical Symposia,* **31:**1–39 (1979).

Bryan, A. C., L. G. Bentivoglio, F. Beerel, H. MacLeish, A. Zidulka, and D. V. Bates: "Factors Affecting Regional Distribution of Ventilation and Perfusion in the Lung," *J. Appl. Physiol.,* **19:**395–402 (1964).

Comroe, J. H.: *Physiology of Respiration,* 2d ed., Year Book, Chicago, 1974, pp. 8–21.

———, R. E. Forster, A. B. Dubois, W. A. Briscoe, and E. Carlsen: *The Lung—Clinical Physiology and Pulmonary Function Tests,* 2d ed., Year Book, Chicago, 1962.

Cotes, J. E.: *Lung Function: Assessment and Application in Medicine,* 4th ed., Blackwell, Oxford, 1979.

Macklem, P. T., and B. R. Murphy: "The Forces Applied to the Lung in Health and Disease," *Am. J. Med.,* **57:**371–377 (1974).

Milic-Emili, J.: "Ventilation," in J. B. West (ed.), *Regional Differences in the Lung,* Academic, New York, 1977, pp. 167–199.

———: "Pulmonary Statics," in J. G. Widdicombe (ed.), *MTP International Review of Sciences: Respiratory Physiology,* Butterworths, London, 1974, pp. 105–137.

Murray, J. F.: *The Normal Lung,* Saunders, Philadelphia, 1976, pp. 89–95, 308–314.

Nunn, J. F.: *Applied Respiratory Physiology,* 2d ed. Butterworths, London, 1977, pp. 213–245.

Otis, A. B.: "Quantitative Relationships in Steady-state Gas Exchange," in W. O. Fenn, and H. Rahn (eds.), *Handbook of Physiology,* vol. 1, sec. 3: "Respiration," American Physiological Society, Washington, D.C., 1964, pp. 681–698.

Pappenheimer, J. R., J. H. Comroe, Jr., A. Cournand, J. K. W. Ferguson, G. F. Filley, W. S. Fowler, J. S. Gray, H. F. Helmholz, Jr., A. B. Otis, H. Rahn, and R. L. Riley: "Standardization of Definitions and Symbols in Respiratory Physiology," *Fed. Proc.,* **9:**602–605 (1950).

Slonim, N. B., and L. H. Hamilton: *Respiratory Physiology,* 3d ed., Mosby, St. Louis, 1976, pp. 40–46.

West, J. B.: *Ventilation/Blood Flow and Gas Exchange,* 3d ed., Blackwell, Oxford, 1977, pp. 41–52.

CHAPTER 4

Barer, G. R., P. Howard, and J. W. Shaw: "Stimulus-Response Curves for the Pulmonary Vascular Bed to Hypoxia and Hypercapnia," *J. Physiol.,* **211:**139–155 (1970).

Benumof, J. L., and E. A. Wahrenbrock: "Blunted Hypoxic Pulmonary Vasoconstriction by Increased Lung Vascular Pressures," *J. Appl. Physiol.,* **38:**846–850 (1975).

Bergofsky, E. H.: "Humoral Control of the Pulmonary Circulation," *Ann. Rev. Physiol.*, **42**:221–233 (1980).

Borst, H. G., M. McGregor, J. L. Whittenberger, and E. Berglund: "Influence of Pulmonary Arterial and Left Atrial Pressure on Pulmonary Vascular Resistance," *Circ. Res.*, **4**:393–399 (1956).

Comroe, J. H.: Physiology of Respiration, 2d ed., Year Book, Chicago, 1974, pp. 142–157.

Culver, B. H., and J. Butler: "Mechanical Influences on the Pulmonary Microcirculation," *Ann. Rev. Physiol.*, **42**:187–198 (1980).

Dawson, C. A., D. J. Grimm, and J. H. Linehan: "Influence of Hypoxia on the Longitudinal Distribution of Pulmonary Vascular Resistance," *J. Appl. Physiol.*, **44**:493–498 (1978).

Downing, S. E., and J. C. Lee: "Nervous Control of the Pulmonary Circulation," *Ann. Rev. of Physiol.*, **42**:199–210 (1980).

Dugard, A., and A. Naimark: "Effect of Hypoxia on Distribution of Pulmonary Blood Flow," *J. Appl. Physiol.*, **23**:663–671 (1967).

Fishman, A. P.: "Hypoxia on the Pulmonary Circulation. How and Where It Acts," *Circ. Res.*, **38**:221–231 (1976).

———: "Vasomotor Regulation of the Pulmonary Circulation," *Ann. Rev. Physiol.*, **42**:211–220 (1980).

Green, J. F.: *Mechanical Concepts in Cardiovascular and Pulmonary Physiology*, Lea & Febiger, Philadelphia, 1977, pp. 55–64.

Harris, P., and D. Heath: *The Human Pulmonary Circulation,* 2d ed., Churchill Livingstone, London, 1977.

Hughes, J. M. B., J. B. Glazier, J. E. Maloney, and J. B. West: "Effect of Lung Volume on the Distribution of Pulmonary Blood Flow in Man," *Respir. Physiol.*, **4**:58–72 (1968).

———, J. B. Glazier, J. E. Maloney, and J. B. West: "Effect of Extraalveolar Vessels on Distribution of Blood Flow in the Dog Lung," *J. of Appl. Physiol.*, **25**:701–712 (1968).

Maloney, J. E., D. H. Bergel, J. B. Glazier, J. M. B. Hughes, and J. B. West: "Transmission of Pulsatile Blood Pressure and Flow through the Isolated Lung," *Circ. Res.*, **23**:11–24 (1968).

Murray, J. F.: *The Normal Lung,* Saunders, Philadelphia, 1976, pp. 33–42, 113–150.

Permutt, S., B. Bromberger-Barnea, and H. N. Bane: "Alveolar Pressure, Pulmonary Venous Pressure, and the Vascular Waterfall, *Medicina Thoracalis* (now *Respiration*), **19**:239–260 (1962).

———, P. Caldini, A. Maseri, W. H. Palmer, T. Sasamori, and K. Zierler: "Recruitment versus Distensibility in the Pulmonary Vascular Bed," in A. P. Fishman and H. H. Hecht (eds.), *The Pulmonary Circulation and Interstitial Space* University of Chicago Press, Chicago, 1969, pp. 375–390.

Szidon, J. P., and A. P. Fishman: "Autonomic Control of the Pulmonary Circulation," in A. P. Fishman and H. H. Hecht (eds.), *The Pulmonary Circulation and Interstitial Space,* University of Chicago Press, Chicago, 1969, pp. 239–268.

West, J. B.: *Ventilation/Blood Flow and Gas Exchange,* 3d ed., Blackwell, Oxford, 1977, pp. 15–32.

______: "Radioactive Methods; and Blood Flow," in J. B. West (ed.), *Regional Differences in the Lung,* Academic, New York, 1977, pp. 33–84, 85–165.

______, C. T. Dollery, and A. Naimark: "Distribution of Blood Flow in Isolated Lung; Relation to Vascular and Alveolar Pressures," *J. Appl. Physiol.,* **19**:713–724 (1964).

CHAPTER 5

Bouhuys, A.: *The Physiology of Breathing,* Grune & Stratton, New York, 1977, pp. 69–74.

______, R. Jonsson, S. Lichtneckert, S. E. Lindell, C. Lundgren, G. Lundin, and T. R. Ringquist: "Effects of Histamine on Pulmonary Ventilation in Man," *Clinical Science,* **19**:79–94 (1960).

Comroe, J. H.: *Physiology of Respiration,* 2d ed., Year Book, Chicago, 1974, pp. 168–182.

Dejours, P. (translated by L. Farhi): *Respiration,* Oxford, New York, 1966, pp. 54–56, 201–212.

Nunn, J. F.: *Applied Respiratory Physiology,* 2d ed., Butterworths, London, 1977, pp. 277–286.

West, J. B.: *Ventilation/Blood Flow and Gas Exchange,* 3d ed., Blackwell, Oxford, 1977*a*, pp. 33–52.

______: *Respiratory Physiology—The Essentials,* 2d ed., Williams & Wilkins, Baltimore, 1979, pp. 51–64.

______: "Gas Exchange," in J. B. West (ed.), *Regional Differences in the Lung.* Academic, New York, 1977*b*, pp. 202–243.

CHAPTER 6

Bates, D. V., P. T. Macklem, and R. V. Christie: *Respiratory Function in Disease,* 2d ed., Saunders, Philadelphia, 1971, pp. 75–92.

Comroe, J. H.: *Physiology of Respiration,* 2d ed., Year Book, Chicago, 1974, pp. 158–167.

______, R. E. Forster, A. B. DuBois, W. A. Briscoe, and E. Carlsen: *The Lung,* 2d ed., Year Book, Chicago, 1962, pp. 111–139.

Cotes, J. E.: *Lung Function,* 4th ed., Blackwell, Oxford, 1979, pp. 203–250.

Forster, R. E.: "Diffusion of Gases," in W. O. Fenn and H. Rahn, (eds.), *Handbook of Physiology,* vol. 1, sec. 3: "Respiration," American Physiological Society, Washington, D.C., 1964, pp. 839–872.

Nunn, J. F.: *Applied Respiratory Physiology,* 2d ed., Butterworths, London, 1977, pp. 310–333.

Wagner, P. D., and J. B. West: "Effects of Diffusion Impairment on O_2 and CO_2 Time Courses in Pulmonary Capillaries," *J. Appl. Physiol.,* **33**:62–71 (1972).

West, J. B.: *Respiratory Physiology, the Essentials,* 2d ed., Williams & Wilkins, Baltimore, 1979, pp. 22–31.

CHAPTER 7

Comroe, J. H.: *Physiology of Respiration,* 2d ed., Year Book, Chicago, 1974, pp. 183–196.
Davenport, H. W.: *The ABC of Acid-Base Chemistry,* 6th ed., The University of Chicago Press, Chicago, 1974, pp. 3–49.
Jacquez, J. A.: *Respiratory Physiology,* McGraw-Hill, New York, 1979, pp. 154–185.
Lambertsen, C. J.: "Transport of Oxygen, Carbon Dioxide, and Inert Gases by the Blood," in V. B. Mountcastle (ed.), *Medical Physiology,* 14th ed., Mosby, St. Louis, 1980, pp. 1721–1748.
Nunn, J. F.: *Applied Respiratory Physiology,* 2d ed., Butterworths, London, 1977, pp. 334–346, 399–412.
Roughton, F. J. W.: "Transport of Oxygen and Carbon Dioxide," in W. O. Fenn and H. Rahn (eds.), *Handbook of Physiology,* vol. 1, sec. 3: "Respiration," The American Physiological Society, Washington, D.C., 1964, pp. 767–825.
Severinghaus, J. W.: "Blood Gas Calculator," *J. Appl. Physiol.,* **21**:1108–1116 (1966).
Slonim, N. B., and L. H. Hamilton: *Respiratory Physiology,* 3d ed., Mosby, St. Louis, 1976, pp. 77–93.

CHAPTER 8

Davenport, H. W.: *The ABC of Acid-Base Chemistry,* 6th ed., The University of Chicago Press, Chicago, 1974.
Gardner, M. L. G.: *Medical Acid-Base Balance: The Basic Principles,* Baillière Tindall, London, 1978.
Huber, Gary L.: *Arterial Blood Gas and Acid-Base Physiology,* A Scope Current Concepts Publication, The Upjohn Company, Kalamazoo, 1978.
Masaro, E. J., and P. D. Siegel: *Acid-Base Regulation: Its Physiology, Pathophysiology and the Interpretation of Blood Gas Analysis,* 2d ed., Saunders, Philadelphia, 1977.
Pitts, R. F.: *Physiology of the Kidney and Body Fluids,* 3d ed., Year Book, Chicago, 1974, pp. 178–241.
Vander, A. J.: *Renal Physiology,* 2d ed., McGraw-Hill, New York, 1980, pp. 137–164.

CHAPTER 9

Berger, A. J., R. A. Mitchell, and J. W. Severinghaus: "Regulation of Respiration," *N. Engl. J. Med.,* **297**:92–97, 138–143, 194–201 (1977).
Cohen, M. I.: "Neurogenesis of Respiratory Rhythm in the Mammal," *Physiol. Rev.,* **59**:1105–1173 (1979).
Comroe, J. H., Jr.: *Physiology of Respiration,* 2d ed., Year Book, Chicago, 1974, pp. 22–93.

Jacquez, J. A.: *Respiratory Physiology,* McGraw-Hill, New York, 1979, pp. 236–350.
Lambertsen, C. A.: "Neural Control of Respiration; Chemical Control of Respiration at Rest; and Physical, Chemical, and Nervous Interactions in Respiratory Control," in V. B. Mountcastle (ed.), *Medical Physiology,* 14th ed., Mosby, St. Louis, 1980, pp. 1749–1827, 1873–1900.
Maren, T. H.: "Cerebrospinal Fluid, Aqueous humor, and Endolymph," in V. B. Mountcastle (ed.), *Medical Physiology,* 14th ed., Mosby, St. Louis, 1980, pp. 1218–1252.
Mitchell, R. A.: "Control of Respiration," in E. D. Frohlich (ed.), *Pathophysiology,* 2d ed., Lippincott, Philadelphia, 1976, pp. 131–147.
———, and A. J. Berger: "Neural Regulation of Respiration," *Am. Rev. Respir. Dis.,* **111**:206–224 (1975).
Murray, J. F.: *The Normal Lung,* Saunders, Philadelphia, 1976, pp. 223–251.
Wasserman, K.: "Breathing During Exercise," *N. Engl. J. Med.,* **298**:780–785 (1978).
Wyman, R. J.: "Neural Generation of the Breathing Rhythm," *Ann. Rev. Physiol.,* **39**:417–448 (1977).

CHAPTER 10

Comroe, J. H.: *Physiology of Respiration,* 2d ed., Year Book, Chicago, 1974, pp. 285–292.
Fishman, A. P.: "Nonrespiratory Functions of the Lung," *Chest,* **72**:84–89 (1977).
Green, G. M., G. J. Jakab, R. B. Low, and G. S. Davis: "Defense Mechanisms of the Respiratory Membrane," *Am. Rev. Respir. Dis.,* **115**:479–514 (1977).
Harris, P., and D. Heath: *The Human Pulmonary Circulation,* 2d ed., Churchill Livingstone, Edinburgh, 1977, pp. 204–207.
Hoffman, J. I. E., A. Guz, A. A. Charlier, and D. E. L. Wilcken: "Stroke Volume in Conscious Dogs; Effect of Respiration, Posture, and Vascular Occlusion," *J. Appl. Physiol.,* **20**:865–877 (1965).
Murray, J. F.: *The Normal Lung,* Saunders, Philadelphia, 1976, pp. 145–148, 277–306.
Newhouse, M., J. Sanchis, and J. Bienenstock: "Lung Defense Mechanisms," *N. Engl. J. Med.,* **295**:990–998, 1045–1052 (1976).
Proctor, D. F.: "Physiology of the Upper Airway," in W. O. Fenn and H. Rahn (eds.), *Handbook of Physiology* vol. 1, sec. 3: "Respiration," American Physiological Society, Washington, D.C., 1964, pp. 309–345.
———: "The Upper Airways," *Am. Rev. Respir. Dis.,* **115**:97–129, 315–342 (1977).
Tierney, D. F.: "Lung Metabolism and Biochemistry," *Ann. Rev. Physiol.,* **36**:209–231 (1974).
Weibel, E. R.: "Design and Structure of the Human Lung," in A. P. Fishman (ed.), *Pulmonary Diseases and Disorders,* vol. 1, McGraw-Hill, New York, 1980, pp. 224–271.

CHAPTER 11

Exercise

Asmussen, E.: "Muscular Exercise," in W. O. Fenn and H. Rahn (eds.), *Handbook of Physiology,* vol. 2, sec. 3: "Respiration," American Physiological Society, Washington, D.C., 1965, pp. 939–978.
Harf, A., T. Pratt, and J. M. B. Hughes: "Regional Distribution of $\dot{V}_A/\dot{Q}$ in Man at Rest and with Exercise Measured with Krypton-81 m.," *J. Appl. Physiol.: Respirat. Environ. Exercise Physiol.,* **44:**115–123 (1978).
Margaria, R., and P. Cerretelli: "The Respiratory System and Exercise," in H. B. Falls (ed.), *Exercise Physiology,* Academic, New York, 1968, pp. 43–78.
Maron, M. B., L. H. Hamilton, and M. G. Maksud: "Alterations in Pulmonary Function Consequent to Competitive Marathon Running," *Med. Sci. Sports,* **11:**244–249 (1979).
Murray, J. F.: *The Normal Lung,* Saunders, Philadelphia, 1976, pp. 253–276.
Wasserman, K., and B. J. Whipp: "Exercise Physiology in Health and Disease," *Am. Rev. Respir. Dis.,* **112:**219–249 (1975).

Altitude

Bisgard, G. E., J. A. Will, I. B. Tyson, L. M. Dayton, R. R. Henderson, and R. F. Grover: "Distribution of Regional Lung Function during Mild Exercise in Residents of 3100 m.," *Respir. Physiol.,* **22:**369–379 (1974).
Guenter, C. A., M. H. Welch, and J. C. Hogg: *Clinical Aspects of Respiratory Physiology,* Lippincott, Philadelphia, 1976, 1977, pp. 3–37.
Lambertsen, C. J.: "Hypoxia, Altitude and Acclimatization," in V. B. Mountcastle (ed.), *Medical Physiology,* 14th ed., Mosby, St. Louis, 1980, pp. 1843–1872.
Lenfant, C., and K. Sullivan: "Adaptation to High Altitude," *N. Engl. J. Med.,* **284:**1298–1309 (1971).

Diving

Begin, R., M. Epstein, M. A. Sackner, R. Levinson, R. Dougherty, and D. Duncan: "Effects of Water Immersion to the Neck on Pulmonary Circulation and Tissue Volume in Man," *J. Appl. Physiol.,* **40:**273–299 (1976).
Bennett, P. B., and D. H. Elliot (eds.): *The Physiology and Medicine of Diving and Compressed Air Work,* 2d ed., Baillière Tindall, London, 1975.
Finley, J. P., J. F. Bonet, and M. B. Waxman: "Autonomic Pathways Responsible for Bradycardia on Facial Immersion," *J. Appl. Physiol.: Respirat. Environ. Exercise Physiol.,* **47:**1218–1222 (1979).
Hills, B. A.: *Decompression Sickness,* vol. 1, Wiley, Chicester, 1977.
Hong, S. K., P. Cerretelli, J. C. Cruz, and H. Rahn: "Mechanics of Respiration during Submersion in Water," *J. Appl. Physiol.,* **27:**535–538 (1969).
Strauss, R. H. (ed.): *Diving Medicine,* Grune & Stratton, New York, 1976.
———: "Diving Medicine," *Am. Rev. Respir. Dis.,* **119:**1001–1023 (1979).

APPENDICES

I. SYMBOLS USED IN RESPIRATORY PHYSIOLOGY

P	Partial pressure of a gas (torr)
V	Volume of a gas (ml)
$\dot{V}$	Flow of gas (ml/min, liter/s)
Q	Volume of blood (ml)
$\dot{Q}$	Blood flow (ml/min)
F	Fractional concentration of a gas
C	Content or concentration of a substance in the blood (ml per 100 ml blood)
S	Saturation in the blood (%)

I	Inspired
E	Expired
A	Alveolar
T	Tidal
D	Dead space
a	Arterial
v	Venous
$\bar{v}$	Mixed venous
c	Capillary
c′	End capillary

II. THE LAWS GOVERNING THE BEHAVIOR OF GASES

1 **Avogadro's hypothesis** For all gases, an equal number of molecules in the same space and at the same temperature will exert the same pressure. (One mole of any gas will contain 6.02×10^{23} molecules and will occupy a volume of 22.4 liters at a temperature of 0°C and a pressure of 760 mmHg.)

2 **Dalton's law** In a gas mixture the pressure exerted by each individual gas in a space is independent of the pressures of other gases in the mixture, e.g.,

$$P_{alv} = P_{H_2O} + P_{O_2} + P_{CO_2} + P_{N_2}$$
$$P_{gas_1} = \% \text{ of total gases} \times P_{total}$$

3 **Boyle's law**

$$P_1V_1 = P_2V_2$$

4 **Charles' law**

$$\frac{V_1}{V_2} = \frac{T_1}{T_2} \quad \text{and} \quad \frac{P_1}{P_2} = \frac{T_1}{T_2}$$

5 **Ideal gas law**

$$PV = nRT$$

6 **Henry's law** The weight of a gas absorbed by a liquid with which it does not combine chemically is directly proportional to the pressure of the gas to which the liquid is exposed (and its solubility in the liquid).

7 **Graham's law** The rate of diffusion of a gas (in the gas phase) is inversely proportional to its molecular weight.

8 **Fick's law of diffusion**

$$\dot{V}_{gas} = \frac{A \times D \times (P_1 - P_2)}{T} \qquad D \propto \frac{\text{solubility}}{\sqrt{\text{molecular weight}}}$$

Index

Index